Writing in SOCIOLOGY

Writing in SOCIOLOGY

Mark Edwards
Oregon State University

Los Angeles | London | New Delhi
Singapore | Washington DC

Los Angeles | London | New Delhi
Singapore | Washington DC

FOR INFORMATION:

SAGE Publications, Inc.
2455 Teller Road
Thousand Oaks, California 91320
E-mail: order@sagepub.com

SAGE Publications Ltd.
1 Oliver's Yard
55 City Road
London, EC1Y 1SP
United Kingdom

SAGE Publications India Pvt. Ltd.
B 1/I 1 Mohan Cooperative Industrial Area
Mathura Road, New Delhi 110 044
India

SAGE Publications Asia-Pacific Pte. Ltd.
33 Pekin Street #02-01
Far East Square
Singapore 048763

Acquisitions Editor: David Repetto
Editorial Assistant: Maggie Stanley
Production Editor: Karen Wiley
Copy Editor: Barbara Corrigan
Typesetter: Hurix Systems Pvt. Ltd.
Proofreader: Gail Fay
Indexer: Jeanne Busemeyer
Cover Designer: Anupama Krishnan
Marketing Manager: Erica DeLuca
Permissions Editor: Karen Ehrmann

Printed in the United States of America.

Library of Congress Cataloging-in-Publication Data

Edwards, Mark Evan.
Writing in sociology/Mark Edwards.

p. cm.
Includes bibliographical references and index.

ISBN 978-1-4129-1424-6 (pbk.)

1. Sociology—Authorship. 2. Sociology—Research.
3. Report writing. 4. Sociology—Study and teaching (Higher) I. Title.

HM569.E39 2012

808'.066301—dc23

2011020082

This book is printed on acid-free paper.

12 13 14 15 10 9 8 7 6 5 4 3 2

Contents

Preface

In the late 1990s, I began my career as a sociology professor at Oregon State University, hired specifically to teach research methods and sociological writing to undergraduates. In the subsequent years, as our department reoriented toward also teaching graduate students, I discovered that many of the things I was saying to undergraduates I was saying to graduate students as well. The challenges of writing confront us throughout our education and our careers. My hope has been for this project to encourage, comfort, and train anyone who wants to effectively communicate sociological insights.

As in most writing projects, there was a community of scholars and friends (not mutually exclusive groups, to be sure) who made it possible. Thanks to Vicki Tolar Burton for her vision and leadership in advancing the Writing Intensive Curriculum program at our school. With a small grant from the Writing Intensive Curriculum at Oregon State University, my colleague Dwaine Plaza and I were assisted in the process of organizing our department's faculty to create a writing handbook for our undergraduates around 2000. In the original version of this handbook my senior colleagues Gary Tiedeman, Loyd Klemke, Sheila Cordray, and Rich Mitchell (all now retired) and my not-so-senior colleagues Dwaine Plaza, Lori Cramer, Denise Lach, Rebecca Warner, and Sally Gallagher contributed chapters that have been valuable to the sociology majors at Oregon State University for the past 10 years.

This significantly revised version of that original handbook has retained some of the original work of Tiedeman, Mitchell, Cordray, and Lach while less obviously folding ideas from the others into the fully rewritten chapters I have created.

Annually teaching a course that requires students to write like sociologists has convinced me that writing is both personal and social, emotional and rational, daunting and yet often gratifying. I recall with satisfaction the students who have written back to me to say how proud they were of their own progress as writers or thankful that they studied in a department that encouraged writing in the discipline. It is for the future cohorts of sociology students that I have prepared this writing handbook. Those who purchase it may wish to know that most of the royalties from the sales of this book provide resources for departmental writing awards, scholarships, and conference fees for sociology students.

Section I

Thinking Broadly about Writing Sociology

Chapter One

Introduction

Shaking his head and smirking a little, Carl asked me, "Do you know what I did when I was ready to turn in my paper for your class?"

"What?" I replied, expecting to hear about some superstitious performance ritual, like many of my students have confessed.

"I showed it to my wife."

"Hmmm. What's the story there?"

"I've always been embarrassed of what a lousy writer I am. I work like crazy on a paper but can never get it to sound right. I never let people read my writing because I know it's weak. Not even my wife. But after working on this paper, with all the revisions and feedback from everyone, I felt like I should let her read my paper."

"What did she think of it?"

"She said I sounded confident."

"Wow. How long have you been married?"

"Five years."

This conversation is telling. Here's a married man who I suspect is intimate with his wife in many ways and yet who would not show her his writing. I hope she recognized what a risk he was taking when he finally let her read it.

Carl's fears about writing betray just how emotional and personal writing can be for us. I suspect there are many who would rather let others see them naked than let others read their writing. Maybe when we let others read our writing we feel naked. Indeed, writing—and letting others read it—is a risky, revealing act. Senior faculty have handed me draft manuscripts, saying in hushed tones, "Clearly it's a work in progress . . ." and then have gone on listing qualifications, self-flagellating in an attempt to influence how I will interpret any weaknesses that are in the text.

Ironically, while writing feels like a very personal performance, it is not merely an individual act of self-revelation. When done well, it is a social act too. In the process of letting others read drafts, give feedback, and talk out loud about our work, we can participate in a community of fellow writers or protowriters seeking to improve our craft. In my classes I tell students to try to think of one another as fellow writers and thinkers, helping each other to improve. But for most, it is difficult to quit thinking of writing as the place where there's no faking it, where others get to see that we don't know as much as we want them to think we know.

Undergraduates and early career graduate students face similar fears of sounding foolish or utterly failing to do what is expected. As a result, rather than just saying, "Aw, pull it together," I try in this handbook to empathetically address the concerns that confront us when writing within the discipline of sociology. Based on my reviews of hundreds of student research papers, talking privately and in class about the writing process, and with help from my friends who have read drafts, I try to demystify the process and help students make progress like Carl did. While it is not possible to give you a sufficiently detailed prescription or recipe for constructing the perfect paper, or at least the paper that will satisfy your professor or thesis committee, it is possible to learn to avoid rookie mistakes and to pick up some tips from those who have gone before you.

The book begins with advice about how to create researchable questions and then provides an overview of a typical academic research paper. It then addresses two of the most confounding issues that students face: how to borrow well from existing research literature and how to document the use of that literature. The middle of the book is arranged in two sections. While some of the best research available draws well from both quantitative and qualitative methodologies, most papers we read in journals and most books tend to fall into one of these two styles of research. So the book first focuses on writing a quantitative research paper, recognizing that many of the things said there apply to qualitative papers. Then subsequent chapters turn to qualitative research, including more detail about writing with qualitative data as well as using these techniques in case studies and internship journals. Finally, the handbook includes other kinds of sociological writing students are asked to complete, including book reviews and theory/content papers. This handbook is not meant to be read front to back like a novel.

But wherever you begin to read, you will find consistency in the advice given and the sense of hope and humility with which it is given.

While a book can provide advice, there is no substitute for reading sociological texts as a way of learning how to write. Most course textbooks themselves don't provide you that insight because they are written for students who are new to the discipline. But reading journals and monographs by sociologists and noting how they say it and what it sounds like when they say it will help you gain an ear for good sociological writing. So I urge you to read a few sociological articles out loud (probably not in the presence of friends or family—they'll think you're nuts) to hear how authors articulate their ideas. And see if the advice given in this book squares up with how published sociologists have constructed their articles and books.

Chapter Two

Turning Ideas into Researchable Questions

For many students, selecting a research question is harder than answering it. Perhaps it's the overwhelming sense that there are so many things to choose from. Or maybe it's the paralysis that comes with the fear of picking a lousy, boring, unanswerable question. We don't want people to eventually laugh at us, saying, "What were you thinking?!" Dr. Edward Witten, thought by some people to be one of the world's greatest physicists, once said this about picking a research question: "You want to find the question that is sufficiently easy that you might be able to answer it, and sufficiently hard that the answer is interesting. You spend a lot of time thinking and you spend a lot of time floundering around" (http://edition.cnn.com/2005/TECH/science/06/27/witten.physics/). It's that floundering around that is so frustrating. Thankfully, some professors mercifully narrow the range of possible topics. And bosses usually tell us which research question they are paying us to answer. But often we are given some freedom to formulate our own questions, and that can be disorienting.

The following material can't make you more decisive, but it can help you figure out how to take vague intuitions and turn them into researchable ideas. Whether you are picking a topic for a class paper or a senior or graduate thesis, these ideas can help you make progress toward articulating a question that can be answered using social science methods. As a graduate student in sociology, I recall spending too many worrisome nights lying there thinking, "The weeks are passing by, and I still don't know what I'm writing about!" The following material is my effort to spare you some similarly anxious nights.

Storks Don't Deliver Them

Where do research ideas come from? Many academics would have you believe that most ideas come from carefully reading the literature. Not true. Almost no one I've ever studied with in grad school, worked with as a professor, or supervised for a thesis or term paper has come up with his or her idea merely by looking for holes in the literature. Sure, sometimes students or other researchers read an article and think, "That can't be right," and decide to verify it, or we see two writers come to conflicting conclusions, thus suggesting a new researchable question. But this is much less common than you might think.

Instead, ideas pop into our heads, fall into our laps, or are thrust upon us. Of course, they may seem like they spring out of the blue, but usually there are circumstances, experiences, and people around us who somehow create the situation that produces inspiration. We can get an idea while driving around, getting ready for school or work, hearing about a current event, talking to a friend or classmate, listening to a lecture. It may not be as shocking as a then-it-hit-me! lightning bolt, but an idea emerges, a bit vague, yet with that immediate promise of being something worth pursuing. How exhilarating is that moment when you say to yourself, "That's worth studying" or "I'll bet that could be studied!"

Inspiration often comes from encountering a problem, perhaps at work, at school, in your family, or when dealing with an organization. You are told to trim your budget, and you have to ask, "Which program is the least effective?" or "How can we cause the least amount of damage?" Or someone you supervise says, "Why aren't people using this cool service that we've made available?" or "Is it wise to have her spend all her time working on that?" Your mom calls and complains about the inefficient and impersonal service she gets from her Internet provider, or your grandpa refuses to read the e-mails you send him. You get into an argument with a student about why certain people do or do not support a new piece of health care legislation. We may not immediately think of these as particularly inspiring situations, but if we stop and think about what is problematic and why, we may find right under our noses questions for which answers may not be immediately obvious and questions that we could investigate. And lo and behold, if we do it right, we may learn something to share with others as a result of our investigation. And probably, no one will laugh.

The focus of this material is to assist you with taking vague notions and ideas that cross your mind and turning them into researchable questions that might yield real insights that we can share with others or that we can use to make important decisions.

From Notions to Research Questions

People commonly talk about their thinking in terms of wrestling. "I'm wrestling with this problem" or "wrestling with a decision." This is a useful analogy because people who know how to wrestle know some standard moves that they can use to bring their opponents into submission. This is what we do with the first ideas that might lead to researchable questions. We wrestle them.

There are probably dozens of mental moves we can make when trying to pin down and press an idea into a researchable question. I'll list and discuss eight that I have seen used and have used myself.

People or Organizations?

Let's say you are sitting in the library and you see a fellow student successfully steal a book or piece of equipment from the building.

Your first reaction might be, "Hey, what the &*%^!" or "That's just not right," and then you text a friend to tell him or her what you saw. (You could call the authorities, of course.) But later you might begin to wonder, "What are the other ways that thieves manage to take things from the library?" That is, maybe you are now interested in the techniques of collegiate thieves.

That might be interesting in and of itself, but if you will think like a sociologist and go bigger, you can ask other interesting questions. By this I mean, try turning your attention to the organizations that have to protect themselves against thieves: in this case libraries. You could as easily ask, "What are the characteristics of libraries that reduce theft?"

In the first example, you were interested in individual people and the techniques they use to steal. In the second one, not as often thought of by those without sociological training, you are now thinking about organizations and how they operate.

There's no guarantee that the latter question is ultimately better, but the mental move of switching from merely thinking about individuals to thinking about organizations opens up interesting possibilities.

Here's another example: You notice Girl Scouts selling cookies outside the grocery store. You wonder, "What makes some Girl Scouts more successful salespersons than other Girl Scouts?" OK, that might be interesting, but take it up a level. For example, why are some Girl Scout troops more successful than others at cookie sales? Now, rather than just thinking about whether this little girl has more sales skills than another, you are asking about the structural characteristics of a set of organizations (i.e., Girl Scout troops). Notice that with this move, instead of thinking about individual characteristics (persistence, cuteness, loudness), you can consider sociological concerns such as how organizations are structured, how they deal with logistics (like getting those cookies delivered), how those organizations fit into the local community, and things like that.

Consider one last example: You learn that certain personal traits affect the likelihood that low-income people will ask for public assistance. You are under the impression that lots of people have studied that. But if you move up to a larger aggregation of people (e.g., counties or states) there may also be important things to explore. For example, what are the characteristics of states that affect the rate of requests for help among low-income people? Now you'd be comparing states that have different rates of people asking for help and other characteristics that might affect that (such as how easy various states make it to ask for help or how different states let people know that there is help to ask for).

Less often we start with an idea about organizations. But it can happen. You hear that wood products companies that emphasize quality in their products are more successful than those that emphasize quantity. Is this also true for individuals who work in wood products factories? That is, are wood products workers who work slower but produce greater quality rewarded better than sloppy but highly productive workers? You can see that if a process that is evident for organizations might not be evident for the people who inhabit them, that can make for a very interesting research question.

The point here is that even if some of the questions we derive are lame and should be discarded, this mental move of considering individuals versus organizations might help us discover often-undiscovered questions.

When We Study People, Are We Interested in What They Do or What They Believe?

You might start with a question—Why are my classmates so evenly split on the question of whether we should approve the new fee to pay for an improved sports facility?

There may indeed be many variables that shape people's opinions. But perhaps it's not just people's opinions that are of interest to us. Perhaps it's what they actually do with regard to this issue. That is, perhaps what is more interesting is the degree to which they differ in their use of the current sports complex or the degree to which they differ in their active campaigning for or against the facility. So, an initial idea about opinions can lead to a potentially profitable research question about actions.

Similarly, imagine that you read that many farmers express resentment against federal regulators who tell them what they can and cannot do on their land. You might naively assume that this means they are often doing harmful things to the land and don't want to get caught. But we have no idea what they actually are doing. It could be that they just don't like the government. You might ask, "To what extent do farmers actually do things on their land that are harmful to the environment?" If there is a possible disconnection between what people say and what they do (or in this case, what we think they may be doing), you may have found a particularly important question to examine.

It can go the other way too. Imagine a bunch of professors sitting alone in their offices during their office hours, wondering why students don't come to see them. Perhaps they naively assume that if they keep switching their office hours to times that may be more convenient for students or they remind students of their availability, then students will come. Alas, the students do not. The behavior (not visiting office hours) is what concerns the professors. But they probably don't have any idea what students believe about office hours. When they complain to each other at a faculty meeting, it's likely they would all say "no" if asked, "Have you ever talked with your students about why they don't come?" Perhaps students don't really believe the invitation is sincere, don't really believe that professors will be there, or don't believe that visiting the professors would be beneficial. In this case, the action is what is observed, but what really needs to be explored may be what students believe about office hours and why.

Notice that your decision about how far to pursue the alternate versus original question will be shaped by what interests you, what research your teacher or supervisor will approve, and what kind of data is or could be available to you. It may be that the alternative question is a dud, but the mental move of switching from looking at behaviors to looking at beliefs, or vice versa, may open up useful research ideas.

Widening or Narrowing the Question

The first germ of an idea that comes to our minds is often too abstract or broad to be researchable on its own. Other times it might be too narrow to be of much interest to anyone. Here's what I mean.

You are volunteering at an after-school program and for the fourth straight day, you are nagging at kids to behave, breaking up fights, and dealing with disrespectful children. You say to yourself, "Why are we doing this, anyway?" That's a profound question in many ways, ranging from "What motivates us volunteers to be here?" to "What is the expressed purpose of this after-school program?" to "What is the actual impact of this program on children?" The last one is a social science, researchable question.

That is, your exasperated notion about whether this program is useful can lead you to a specific question: Which student outcomes from this program can be measured? Sometimes an exasperated question of frustration (probably too wide and vague) can generate a useful, narrow, specific, researchable question.

Consider this one: A Parent-Teacher Association (PTA) president claims, "Students should be required to wear uniforms." Statements and questions that include *should* are not researchable on their own, but they probably imply researchable questions. Sometimes *should* implies moral claims about what fits within acceptable behavior. This president might mean that because she values modesty or formality, the students ought to comply with norms that emphasize modesty and formality. (This is what sociologists call a "normative claim" because it focuses on getting people to abide by social norms.) With a little bit of thinking, you might derive from this sentence a question such as, "Which intentional and unintentional messages do public school students send with the clothes they wear?" With that we might be able to consider the degree to which students and nonstudents agree on the meaning of their attire, perhaps using data obtained from surveys of both groups.

Alternatively, *should* may imply what is wisest or what is most efficient. This same PTA president may have meant her statement to be a practical solution to a cause-effect process she thinks is going on. Perhaps she is theorizing that school uniforms will cut down on distracting behaviors and increase student performance. If that's what she meant, then that's something we can study.

Conversely, sometimes a notion will come to us, and immediately it is obvious that it's too narrow. For example, imagine that you say to yourself, "Why does my neighbor insist on filling his yard with pink flamingos?" His behavior may be curious, but this is of course too narrow. But asking, "What is this an instance of?" can help me generate a broader, more interesting question. If pink flamingos are a case of self-expression through one's yard, then this move suggests larger questions about how people use their homes (gardens, house, yard, etc.) as a way of expressing themselves. We could then ask, "What are the categories of self-expression that we see in people's homes?" and "Why do some people express themselves one way versus another through how they decorate their yards?"

Here's one last example, again about self-expression. Your friend comes over and says, "What do you think of this tattoo of our school mascot now covering the left side of my face?" First piece of advice: As a friend, tell her it looks great even if internally you are saying, "What the heck was she thinking?!" But that question about this one person is of course too narrow for a sociologist. As a sociologist your researchable question might be, "What characteristics of people influence whether they get big tattoos on their faces?" Even this might seem too small to you, and you may wish to expand this to examine how other forms of self-expression (other tattoos, body piercings, body modifications, unusual haircuts, loud Hawaiian shirts, etc.) are shaped by various personal and social variables.

Again, the resulting questions you come up with may not be perfect, but in the effort to widen and narrow questions, you can mentally play with the topic in a way that can lead to the articulation of questions you might otherwise never think of. You may also discover that by widening and narrowing your question you can link your interests to those of other researchers. Others may not be writing about facial tattoos or pink flamingos, but their work and yours may have much in common.

Descriptive Versus Causal Questions

Sometimes an idea will come that is descriptive in nature. For example, you may observe students wandering around the library apparently looking for a quiet place to study. It may appear to you that there are not enough quiet places for students to study. You could decide to then count the number of students and the number of quiet places in the library, but that isn't very interesting (unless you are a librarian in charge of quiet spaces). But if I think about turning this descriptive project (describing the ratio of quiet places to students) into a causal question, I can start to ask questions such as the following:

a. Which changes in library policies have affected the number of quiet places available to study in the library?
b. What has caused library administrators to care less about keeping libraries quiet than in the past?
c. What is the role of student demand in shaping library quiet policies?

Notice that all three of those may be lame questions or really interesting ones. There's no guarantee that you'll immediately discover a good one, but by moving from descriptions to causal questions, you open up new opportunities for interesting ideas.

Here's another example: "We are wondering how many people in our church congregation would resent it if we got rid of the wooden pews." It could be that the church leaders asking this question might have in mind that if only a small minority of people would be upset, then it would cause little problem to proceed with removing the pews. But if they start thinking about who among them is most likely to be upset, they may really be interested in the degree of correlation between age and appreciation for pews. That is, they might want to know if the older people would be especially upset or the younger people especially relieved. They may or may not care too much about how age causes appreciation for pews, but we social scientists might care about that. So if there's a simple descriptive question that first emerges, you can sometimes locate an interesting causal, or at least correlational, question to consider.

This switch from descriptive to causal is not merely the act of inserting new variables. Sometimes the variables are readily apparent if we just listen to how we say things. For example, a student recently started with this statement: "I'm interested in understanding the high level of poverty among single American women." That's an excellent place to start a study. But it does of course imply that she knows how to judge what a high level is, and that presumably that level is different for single women compared to some other group (single men? married women? who?). So although her first interest is in this problem among one group, really she is interested in comparing the poverty rate for single women versus some other group, and perhaps she can find a way to explain these differences. And in making the comparison, she will be able at least to say, "It's higher for this group than for that group." Making that simple move of clarifying the implied comparison immediately reveals that she can compare at least two groups (single women and others), but it also will help her begin to imagine other variables that are likely to matter (men's vs. women's pay rates, married vs. unmarried persons' pay rates, average number of hours worked, occupational prestige, etc.). So being clearer about the implied comparison helps her to see the other interesting variables that probably affect the likelihood that anyone would be poor in the United States.

Sometimes descriptive questions turn into causal questions once we start studying the topic. That is, your question may morph into something different once you begin. For example, some years back I was interested in understanding pregnancy discrimination in the workplace and describing how the courts had dealt with that issue. After reading through 50 appeals court cases I began to observe that the women suing their employers came from a variety of occupations, some blue collar, others white collar. This led me to hypothesize that there might be a difference in the kinds of claims and concerns raised by those in one kind of occupation versus another. In this case, the question began as a description and turned into a causal analysis about how occupations shape the kinds of complaints we make about work-family concerns.

Admittedly, sometimes our first notions are causal, but they beg for important descriptive questions to be answered. For example, "How does education affect support for spending on state government programs?" is a causal question, but it might inspire a simpler question, "What have been the trends over time in public support for state government programs, and what is the current level of support?" The first question is more common in a sociological paper, but the second one might be exactly what you need to know as you lobby people to vote for or against candidates who have expressed their opinions about state government programs.

Examining Alternative Explanations

Patterns sometimes jump out at us. Rich people spend Friday nights at the opera, and working-class people spend them at the monster truck pull. Rural communities are more vulnerable to job loss during a recession than are urban areas. Cohabiting couples are more likely to get divorced than couples that never cohabitated. These are all correlations between variables and may imply causal relationships.

But as soon as we think that one thing causes another (e.g., wealth increases tastes for opera, rurality increases economic vulnerability, cohabitation increases chances of divorce) we should ask if there might be alternative causal relationships that could be explored. Is it wealth that increases tastes for opera (perhaps because people can afford opera tickets), or is it really education that increases wealth and tastes for opera? Is it rurality that increases vulnerability, or is it the kinds of industries present in rural versus urban places? Is it cohabitation that increases divorce, or perhaps spouses' nontraditional ideas about marriage that both increase willingness to cohabitate and increase willingness to abandon difficult marriages?

It may turn out that the alternative explanation is the whole explanation or that it merely complicates and illuminates the first one. But the effort to take first appearances of causal relationships and challenge them with measurable competing explanations is a classic, common move by sociologists looking for the really important questions to answer.

Quantitative Data, Qualitative Data, or Both?

Another move that can generate interesting researchable questions is to ask, "Which kind of data can we get?" Typically your advisor will tell you, "Get the data that will answer your question." That's true, but before you think about which data are possible to get, or which kind of data you want to spend your time studying, you can use this concern about data to push you to think about alternative questions that can derive from your first notions.

Let's say you read that kids who are bullied are more likely than other kids to join gangs. This initially sounds like a study that relies on quantitative data, perhaps locating a correlation between two variables (getting picked on and joining gangs). But saying to yourself, "If I could personally interview and observe young people, what related things could I examine?"

may open up new ideas such as, "How does having nongang friends help insulate kids who are bullied from joining a gang?" or "How do kids who join gangs frame their stories about entering gang life, and are their stories consistent when telling them to people inside and outside of gangs?" You can see that by just saying, "What could I learn if I had greater personal access?" you might generate interesting researchable ideas.

Similarly, if your first idea focuses on qualitative data, consider quantitative studies that your first notion could inspire. If you are reading a study about the housing difficulties of 10 low-income families, consider what counting might tell us. That is, does anyone know how often poor people move and, if they move, where they move? Might we benefit from knowing what percentage of low-income people move to areas that are better or worse off than the areas they currently live in? In this case, the story describing the processes at work in the lives of families might suggest to us the benefit of merely counting the number of such families and finding out which concerns, problems, and opportunities are more or less common.

How Is Time Relevant to My Question?

Thinking about how time is implicated in your first idea might help you develop additional research questions. Let's say you first think of, or read about, a correlation: "Does driving a hybrid versus gas-burning car correlate with people's convictions about global climate change?" That's a cross-sectional analysis that one could conduct on any given day, asking people what they drive and what they believe. But if you follow people's driving habits over time and their beliefs over time, you have a longitudinal study that also is interesting: "How do changes in people's ideas about what's happening to the climate influence their driving decisions and vice versa?" That's a different, and perhaps uniquely interesting, question. If we think about historical sweeping changes, we could also move up to a larger aggregation of people and ask how changes in public opinion about climate and changes in sales of hybrid vehicles are or are not correlated over time.

It works the other way too. If you hear that over time consumption of high-priced coffee in America has increased at the same time that income inequality has increased, you might assume that the upper class has particularly used its money to buy coffee. But this study of parallel historical trends might suggest to us longitudinal studies (how do changes in people's income situation affect the kind of coffee they drink?), or more simply, what percentage of people in different income brackets today drink high-priced coffee?

Letting your mind play with the role of time might guide you into different, otherwise undiscovered questions that you will find more interesting and researchable.

Do I Really Know What I Want to Measure?

This last move often seems like the sort of issue you'd address only once you had picked your question. If you look at research papers you would certainly get that impression since authors talk in the Data and Methods section about measurement. But as you are pondering how to develop your first notions into research questions, pay attention to the words in your initial notions to see if they indicate possible interesting unarticulated concepts. This move is related to the widening/narrowing move, but here the move emphasizes measurement as a way to generate clearer, more researchable ideas.

Consider these questions:

1. Would it be more efficient to sign up volunteers this way or that way?
2. How user-friendly is the university's new website?
3. Does this program work?

Each of these initial ideas could immediately start an interesting debate among friends or colleagues. But inevitably someone will or should say, "What do you mean by . . . ?" Let's consider those three questions to see how this move can help.

1. The person wondering about efficiency should be pressed to answer, "What is efficiency?" and when he or she says, "It means getting as many volunteers signed up as possible per hour of paid staff effort," this suggests new ideas. Indeed, if we take that measurement at face value, perhaps we could see if technique A beats technique B in terms of volunteers per hour of paid staff effort. But in the endeavor to measure it this way, we might discover other interesting questions, such as, "To what extent is productivity (achievement per hour of effort) important in this kind of organization versus a different kind of organization?" or "What are the various ways that nonprofit groups measure and value efficiency in their organizations?"

2. The user-friendliness of something might seem obvious to us because of our experiences with badly designed websites. But when pressed to define it, we may realize that user-friendliness is more complicated than we had thought, in large part because how friendly a website is probably depends on the user. Recognizing the various dimensions of the variable, and thinking about how we would measure it, may inspire us with new ideas such

as, "How do different characteristics of computer users affect the degree to which they can navigate the new website?" Websites that some find friendly are decidedly unfriendly to others. So in this example, the concept of user-friendly begs the question of who the users are and what characteristics of the users, as well as the websites, may be important.

3. We do not know which programs work if we cannot operationalize (i.e., come up with a reasonable measure for) what it means for them to work. The necessity to develop metrics for measuring how well something works may generate uniquely interesting questions. For example, the Department of Human Services wants to know if a marketing campaign for food stamps for low-income people works. By "works," they mean that they are hoping for a certain outcome, such as enrollment in the program. The admission that it is enrollment in the program might suggest a new set of questions: Does hearing about the program increase people's interest in applying for it? Does hearing about this program increase the number of ineligible people applying? If so, by how much?

The mere act of forcing oneself to be clear about measurement can spark new researchable ideas that would otherwise be obscured.

Conclusion

The selection of a research question is rooted in social circumstances and shaped by personal characteristics. Your boss or teacher, your progress toward completion of your degree, the time of the academic term, the availability of data, and your own interests and skills all shape what you'll finally choose. Some of these things you can change or resist, and other things are beyond your control.

However, no matter your circumstances, you can exert some creativity by trying out some of these mental moves to take first ideas, notions, and impressions and turn them around to see what sorts of things emerge. Again, some of the mental moves we have available to us might fail to give us questions we want to answer, but where one may fail, another might succeed. One cannot rationalize creativity with a formula that guarantees inspiration, but working with these mental moves can open up new possibilities.

Chapter Three

Overview of Writing a Research Paper

An Extended Analogy

Imagine that you are a lawyer in court and you need to demonstrate that an employer has been systematically discriminating against older-than-average applicants for jobs. You have a box of applications from potential employees who have applied for jobs during the past year. You also have lists of who was hired, and you have 100 completed questionnaires filled out by managers who were involved in hiring for the company. Interviews with some former employees indicate that their testimony will be useful as well. Now the older applicants have retained you to challenge the employer in court over this topic. Having accepted the case, you now have to figure out how to communicate to the panel of jurors who know little about the problem (age discrimination and the law) that there is good evidence (the box of applications, the lists of people hired, the survey of managers, the testimony of witnesses), and that unfair hiring practices have been used by the employer.

Your lawyerly task of convincing a jury to reach a certain verdict is very similar to the task you face when you are writing an academic research paper. How do you construct a case that will convince the jury? The answer to this question is analogous to how you develop a convincing argument and make a case within an academic research paper. Let's follow the analogy through from beginning to end.

The Introduction

When you write a research paper, you are not writing just for the judge (the teacher). You are writing for an imagined or real audience of peers or a public that knows less about the subject than you do. This puts upon you the task of clearly introducing and explaining the issue before proceeding with the evidence. If you were in court, you would not assume that the jury understood the details of the law or the subtle dynamics of employment discrimination. You would introduce to them the fact that such a phenomenon exists, that there are rules that employers are supposed to follow, and that there is now disagreement between the employer and the unhired applicants regarding whether mistreatment occurred in this instance. You would also make it clear from the start that you intend to convince them that the unhired applicants are right and that the employer is wrong.

Similarly, you begin an academic research paper with an introduction. The introduction alerts the reader to the fact that there is an important phenomenon worthy of our attention and that there is some kind of sociological question surrounding that phenomenon. In addition to alerting the reader to the existence and importance of the topic of your paper, you also alert the reader to the nature of your conclusion. For example, if you were writing a social science paper about age discrimination, instead of proving it in court, you might say, "This study will demonstrate that age discrimination is more common under particular conditions," or something like that.

The Literature Review

After introducing to the jury your intentions and your goals for this case, you then have a chance to explain to the jurors what they need to know about hiring practices, the law, and different ways of understanding this issue. For example, you might want them to know that the law is very specific about age discrimination and that earlier jury cases just like this one have found that employers must be held accountable if they fail to hire someone just because of his or her age. You may want to alert the jurors to their own biases, pointing out that they might tend to feel sympathy for the employer because some of them think that employers should be able to hire whomever they want or that the law has no place in telling employers what to do. You might also think they need to know that this employer is very powerful and wealthy and that it would not have been a major hardship for the employer to have accommodated some of the elderly applicants who applied for jobs. In other words, as the attorney for the unhired applicants, your job is to inform the jurors, review for them the important issues, help them understand what the question really is, and prepare them to carefully judge for themselves.

The literature review of a research paper seeks to accomplish these same tasks. You must have in mind that most of your readers know little or nothing about your topic, and thus you have to review for them the basic features of what is already known and established about this topic. At the same time, some very informed readers (your teacher and your fellow student writers) will be able to check on your accuracy and your honest portrayal of the current state of knowledge. For a further discussion of how to write a good literature review, you should examine the several chapters in this handbook devoted to literature reviews and borrowing from the literature. Remember, the literature review is not usually a full scan of all of the literature. It is a selective but fair treatment of the state of current knowledge about a topic that is designed to point out what is known and what remains to be discovered about a particular social phenomenon and that justifies the reasonableness and importance of the question you are trying to answer.

The Data and Methods Section

If you had a box full of applications from the past year, and 50 percent of them were from older-than-average applicants, but only 2 percent of the new hires were older than average, you could point to the disparity in these percentages as circumstantial evidence that discrimination occurred. But before you could present these findings, you would need to introduce to the jury the fact that you have some data that bear on this issue. Before showing them the 50 percent versus 2 percent gap, you would need to tell the jury about your data. For example, where did you get these records? How reliable are they? Are there missing records that we do not know about? Were these records obtained legally? You also might have to explain how you computed your statistics if you have jurors who do not understand your math. And what about those witnesses? Where did you find them, and what is their credibility for informing you about what was going on at the company?

Similarly, in the Data and Methods section of a research paper you need to tell your readers about your data, how you gathered them, and sometimes how you analyzed them. Some of the questions that may need to be addressed are the following: What is the source of the data? That is, did you collect them? If so, how? Were they collected by the Census Bureau, a private government think tank, other academics, or another source? Did you interview people at length or do a quick survey? Were they derived from telephone surveys, coding of government documents, door-to-door interviews, or another source? How representative of the whole population is your sample? How are you measuring the variables that are in your study? For example, is age measured by "years of age" or by "older than 49/younger than 50"?

There are three main issues to cover in the Data and Methods section: the sample (i.e., the data), the measures, and the strategy for analysis. Put differently, this section of the paper familiarizes the reader with who, what, and how: Who are you including in your study (so that the reader can make a judgment about how much to generalize)? What data and information have you received from your sample? and How will you handle these data?

Findings (or Results)

At some point you introduce to the jury the critical pieces of evidence that demonstrate that the employer engaged in age discrimination. For example, you might show the jurors a pie chart that shows 50 percent of the applications were from people older than 49 and then show a second pie chart that shows that 2 percent of the new hires were older than 49. Then you might introduce to them a table showing that 75 percent of the managers indicated that they believed that older people would be more difficult to work with than younger ones. In other words, you show the results of your analysis that will convince the jury of your claim. At some point you will introduce your key witnesses who have important things to say that will make clear how the managers were talking about various potential employees.

In academic writing, you also will highlight the critical findings of your analysis to point out that the conclusions you will draw are the most reasonable. However, in contrast to a legal case wherein you would hope that any conflicting information be suppressed or overlooked, academic ethics require that you present the whole story, or at least as much as you can. This means that when you complete your analysis, you also report unanticipated or contradictory findings. Then do the best that you can to make sense of these as well. Thus, you should seek to strike a tone of confident assertion while at the same time acknowledging the parts of your analysis that do not support your claim or that might support alternative claims. See the section "Writing Quantitative Papers" for more help with doing this for quantitative papers and the chapter "Ethnographic Interviewing and Storytelling" for qualitative papers.

The Results section of your paper will generally be the place where tables and graphs are located (for quantitative papers) and where quotes and analyzed field notes appear (for qualitative papers).

Discussion and/or Conclusions

In a trial case, the closing argument is the place where you put together the pieces and where you review for people what it is that you have presented to them. You remind them of the legal questions involved, highlight the most critical evidence, point out why you have demonstrated that the alternative claim is false, and suggest to them that they should now decide to agree with you about the guilt of the employer.

Depending on your paper, and your audience, there may be one or two sections that cover the discussion and conclusion. That is, for some papers, the Discussion and the Conclusion sections are the same thing. Why? Sometimes the material does not lend itself first to a discussion of the relevant theoretical issues raised or the surprising findings and then to a more lofty and repetitive section that tells us why this research is so important. But for other papers, there really is something different to be said first about interpreting the results (the Discussion section) and then about making sense of them in more global or expansive terms (the Conclusion section). How you choose to construct the paper after the reporting of findings is a judgment call on your part, as the author, and on the part of your editor (or instructor).

Either way, the last part of the paper is the place where you quickly summarize what you have accomplished, highlighting the major theoretical question (or questions), reminding the reader of the central findings that help answer the question, and pointing out how your explanation is superior to alternatives. You also may need to make sense of weak or insignificant results as well as to suggest potential research that should follow your research and perhaps some of the policy implications of your findings. You need not accomplish all of these things, but at least the restatement and theoretical import of your research must be made clear here. Your conclusion should be concise but also complete enough that if someone read only your conclusion, they would know which question you try to answer, which main findings you provide to answer the question, and what you think your answer ultimately means.

Citations (or References or Bibliography)

Lawyers always need to be ready to cite where they have found legal precedent for the claims they make. When you argue that the employer is guilty you may need to say to the judge or the jury that in *Jones v. Wonka, Inc.*, the judge allowed evidence just like you are providing, and the jury found it convincing enough to convict Wonka, Inc.

In the same fashion, academic writing requires that you indicate where you see in the literature the theoretical or empirical claims that you are evaluating. This section is not just a legally or ethically required component of a paper but is essential for convincing the reader that you have some idea of where your paper fits into the conversation that is going on among academics in the literature. You can learn more about citations and the reference section in the chapter devoted to that topic.

Chapter Four

Borrowing Well from the Literature

One of the most challenging tasks of a sociological research paper or report is situating the project in the existing literature. Sometimes we use the literature to justify trying to answer the question; other times we use it to defend our hypotheses. Sometimes we include literature to merely acquaint the reader with the breadth or narrowness of what has been studied. This variety of reasons for including citations makes clear that we should often ask ourselves why we are including each item, being clear about which task it performs for us.

Whether you are writing a quantitative or a qualitative paper, or a paper that draws on both methodologies, you will be expected to articulate how your paper relates to existing research. This usually happens in the literature review. But before we consider how to organize a literature review, it is important to reflect more broadly on the role you play as a writer who is either more familiar with the existing research than is your audience or who needs to show a sophisticated audience that you yourself are well acquainted with how your work links to what already exists.

I have found it useful to think about the research literature as an ongoing conversation between writers. Like most conversations, it is not scripted nor particularly well organized, and sometimes people are not even listening to each other. So let's work with this second extended analogy, not of a courtroom where you are talking to some jurors but of a room full of people talking about a particular subject. The situation before us challenges us with the question: How do we recount and contribute to the conversation?

Talk, Talk, Talk

Imagine you are in a large room full of hundreds of people, all sitting and standing around in little groups, talking about the topic of employment discrimination. One group is discussing why employers discriminate, another group is discussing which groups are affected, and still another group is puzzling over how to define discrimination. Some of the people standing alone are also saying clever things about these topics, but for whatever reasons, no one is listening to them or no one has figured out how to respond to their ideas. We'll imagine them standing alone in different places apparently muttering to themselves. Let's say that you are particularly interested in age discrimination against workers. So you decide to join one of those groups or start one of your own, and soon you are discussing how we can know when age discrimination is taking place and who is most likely to experience it.

Now, imagine that you were asked later to describe how the things you had to say were related to the other things being discussed in that room. This is one of the intellectual tasks you face as a sociological writer—to map out the state of the conversation and how your part contributes to that conversation.

The good news is that you need not describe every single conversation in that room. You might decide to ignore the tiny conversation in the corner where they were discussing employment discrimination in ancient Rome or the conversation on the other side of the room where they discussed the history of child labor laws. Because you are writing and talking about age discrimination you will need to figure out which conversations (and lone voices) relate to that issue. So you would probably want to explain to someone how your conversation about age discrimination was related to other conversations such as the nearby one about measuring other forms of discrimination or other conversations that were recounting recent trends in age discrimination. To accomplish this task, you need to know what was happening in other conversations, and who was saying what, and then make the judgment about which ones were most relevant to your conversation. Armed with that information, you would be able to say how your little discussion group contributed to the overall conversation in that room.

Notice in this analogy that you need to review not all of the literature but only the literature that is most relevant to your project. Of course you may have to scan widely to see which kinds of conversations are more or less relevant. Some conversations in the literature have been going on for a while, so you might point to some older research projects that initiated this line of discovery, while other items in the research literature are newer. Some topics have yet to be discussed, producing a hole in the literature that new research projects may fill. With the help of colleagues, experts, and supervisors you can decide how much of that conversation needs to be summarized so that readers can see how your project is related to projects already finished.

It is likely that your contribution to this imagined conversation in the literature is pretty small, whether you are a beginner or an accomplished researcher. Aspiring sociology students are often urged to find the hole in the literature that their papers can fill. That's a worthy goal. But it often takes a very long time to scout out such holes, and if you find them, it's unusual that you are in a place to fill such holes. Social science moves forward slowly, and many of us produce research papers that verify and replicate previous findings, or they add only a small new idea. Our term papers—or our multi-thousand-dollar grant proposals—rarely fill a hole in the literature, but we can contribute to conversations in the literature. Imagine in a conversation that someone says, "I discovered blah-ba-de-blah blah," and you say, "Hey, me too!" You have contributed to the conversation. Hence, small projects such as term papers and college theses can confirm expected findings and add to this imagined conversation, if only to say, "I've confirmed others' findings." A small grant proposal to test out a new program or answer a simple question can contribute to the conversation too. When we're lucky, we have a big impact on the conversation, but that's not as common as we might wish.

These reflections about your role in recounting a complex set of conversations do not yet address some practical issues about how much information from each article, book, or report to include and how to include it.

Effectively Recounting the Conversation

If someone asked you to summarize the conversation that was going on in that big room, you would have two options—one bad, one good. One approach would directly quote individuals in each group, and the other would summarize the conversation in your words. The summarizing approach is what literature reviews do well. The quoting approach is what middle schoolers do often, and badly:

> Parent: So, dear, tell me about the bus ride on the field trip with Tyler and Emily.
>
> Middle school child: Well, Tyler said to Emily, "You're, like, so lame." And Emily said, "Oh yeah, well your Facebook page sucks." And Tyler said, "At least I don't have food between my teeth." And Emily said, . . .

The parent probably wants to hear, "Tyler and Emily strongly disagreed about who was least popular and who had poor hygiene."

It's this second approach we are trying to achieve when borrowing from the literature. Here are two examples that parallel the middle schooler example:

> Jones (2001) says, "Age discrimination in employment is the scourge of our country, accounting for untold levels of misery and catastrophic loss." But Smith (2001) says, "Yet the data are unclear as to how common and widespread is the phenomenon of age discrimination." Jones counters this with his claim that . . .

Notice that the author, wishing to acquaint us with Jones and Smith, relies on extensive direct quotes. Indeed, the quoted persons are given credit, but they are quoted word for word, and the resulting text looks like a patchwork of quotes rather than a synthesis and/or summary.

But here's an alternative:

> Jones (2001) and Smith (2001) present conflicting interpretations of the data. Jones argues that age discrimination is widespread and serious, while Smith warns of overstating the case because of weaknesses in existing data sources.

Notice that readers are still acquainted with the idea that there is a conflict in ideas between Jones and Smith, but they are not burdened with the details. And in setting it up this way, the writer asserts authority as the commentator who says, "Here's the status of the conversation between Jones and Smith." That's the creative and authoritative contribution that you, the writer, can make in summarizing the literature.

Now, thinking further about summarizing the hypothetical set of multiple conversations regarding employment discrimination, let's consider our options. One option, often naively embraced by beginning sociological writers, is to mentally go around the room and restate what each person said. This often takes the form of a paragraph per article or book. This style is technically known as an annotated bibliography. That is not a literature review. It is not what you will find in the polished literature reviews of books and articles in sociology. It is a trap into which rookie sociology writers, of both the graduate and undergraduate kind, often fall.

Another option, at the other extreme, is to gloss over any details about who said what and just summarize in abstract terms something about the tone or tenor of the conversation. This sort of approach is attractive if you have not taken the time to listen to any of the conversations (i.e., not read very many articles on the subject), but it will get you in trouble with the discerning reader. Here's an example:

> Scholars disagree about how real and important employment discrimination is. Not much is known about the processes by which age discrimination takes place, but most scholars agree that it is still a problem. Many think that we can learn from Marx's critique of capitalism, while others focus attention on theories about gender socialization and the persistence of gendered expectations.

That sounds very sociological but leaves readers wondering if you just made it up (like I just did!). Making sweeping claims about what many or few or no one has studied or about how much or little is known about a topic, and doing so without citations to support the claims, makes you vulnerable to reasonable criticism by readers who know that in fact you are wrong. Veteran and rookie writers alike fall into this trap when time is short and their reading is cursory.

So the goal is to synthesize an orderly conversation out of the items you have read (avoiding the first trap), having actually done some reading, rather than to make sweeping, unsubstantiated claims (avoiding the second trap). If you again imagine in your mind's eye that room full of small conversations, then you can ask yourself how best to map out that room and be willing to move groups around whose conversations were quite similar. That is, you don't need to say, "Whozit said X over here, but Whatsit said Y over there." You can give readers a sense of how the content of the conversations can be assembled in a way that makes sense, mentally putting people together who really should have been in the same group.

That means that you have permission to summarize what you heard from the different groups, perhaps even blending together some of the groups that thought they were unique but in fact were having the same conversation. That is, maybe two groups were talking separately, one about racial employment discrimination and the other about gender employment discrimination. You could simply point out that research into employment discrimination has focused on gender and racial issues and summarize what they had in common rather than give an exhaustive recounting of each of the two groups. In doing that, you are again asserting creativity and authority.

So here would be a hypothetical summary of the conversation, now no longer thought of as a big room full of discussants but as a discussion in the literature:

> Evidence of age discrimination in employment is not as readily available as one might imagine. Whozit (2001) identifies several legal, organizational, and methodological reasons that researchers have been forced to use indirect methods of measurement. Age discrimination studies have also suffered from inadequate data sources, although greater progress has been made on understanding . . . (Whatzit 2002). But researchers agree that the organizational processes that lead to employment discrimination are similar (Wilson 1999; Elson 2004; Burstein 2005).

Notice now that in this text, the writer helps readers hear the agreement and contradiction among authors and also points out to readers that what is known might contradict readers' prior assumptions. The conversations are not recounted word for word, yet sources are cited to demonstrate that the writer is not just making things up.

As you read research articles, I urge you to look beyond the substance of writing. Watch to see how other authors incorporate research in their writing. You will see various techniques and strategies, some of which I have outlined here and still others you will discover.

Chapter Five

Citing Sources

Why, When, and How

All academic writing requires that the author indicate where in the existing research literature he or she found the theoretical or empirical claims being used in the paper. This requirement is a demand of professional ethics. But by complying with it, you also assure the reader that you know where your paper fits into the conversation that is going on among academics in the literature. In other words, by avoiding plagiarism (stealing other people's ideas and insights), you accomplish something important (helping readers see the contribution of your work and establishing your credibility). Moreover, a thorough reference section allows your readers to investigate the degree to which your paper accurately reports what other authors have said. Unlike newspaper writers, who occasionally cite unnamed sources in news articles, academic writers must divulge the sources of their material. In your research paper, this means that you should include citations for the research papers and books that have provided concepts, claims, or data that are relevant to the research question you are addressing. You may not realize that seasoned academics often skim the bibliography of a paper before they read the paper itself. But researchers often do this because the reference section gives them a clue about how well the author has logically framed the paper and where it stands in relation to other research.

Often, threatened with the big consequences that accompany plagiarism, students raise two important questions about citing sources. First, do I have to find citations to support what everyone already knows? Second, can I cite other people's citations?

Author's Note: Thanks to Sally Gallagher and Dwaine Plaza for earlier ideas and help on this chapter.

Some common sense and some guidance from the American Sociological Association (ASA) can help us out. As expected, the ASA suggests that when you use an author's (or authors') ideas, information, or descriptions within your paper it is required that you make sure that the author is properly cited within the text of your paper. But you don't need to go find citations for things that one could easily assert and assume the readers will agree with. For example, if you say, "Conservatives distrust big government," you do not need a reference to support that because everyone knows that this is in part a defining characteristic of what it means to be conservative. Or if you wish to assert that twentieth-century black Americans were denied many civil rights until federal legislation in 1964, this is widely accepted as historical fact. However, if you wanted to argue that conservatives are twice as likely as liberals to vote for small government or that black Americans voted at half the rate of white Americans in 1955, you would need to cite someone else's empirical evidence because you are making an empirical claim. So when you make empirical claims, you must tell the reader where those claims come from; hence, you use citations.

With regard to the second question about citing other people's citations, the answer is, "Well, if you have to." Sometimes you will find a research article that alerts you to other research articles useful for your study. For two reasons it would be best if you located and read the original. First, if you read the original and find it useful, then you can cite it directly because you have examined it yourself. Second, if you don't read the original, you would not know if the author who alerted you to it misquoted or misrepresented the ideas of the original. But if you cannot obtain the original, then you can indicate what the original paper allegedly argues, according to the author whose paper you did read. Let's say you read Smith's (2010) article wherein he comments on Wilson: ". . . but Wilson's (1995) study found ample evidence of education's effect on . . ." It would be unethical to insert Wilson's (1995) study into your paper as if you had studied it and are now reporting it. It would be better for you to say, "Smith (2010) points out that an early study by Wilson (1995) also demonstrates that education had an impact on . . ." In the reference section (bibliography), you'd list Wilson but with Smith's locating information. (See later discussion.)

These kinds of situations occur most often when one is in a hurry, perhaps because of procrastination, and one doesn't think he or she has time to read the originals. So one can often avoid the problem by planning and reading ahead. If you are unsure about how to cite items mentioned by others but need to do it, then consult a more thorough style guide and talk with the community of writers of which you are a part—your fellow students, mentors, and professors. Whatever you decide about how and when to cite sources introduced by others, at least remember this: It is when we seek to deceive the reader that we cross the line into unethical behavior.

Sometimes we learn of a research paper in the popular press, on the Internet, or through word of mouth. News magazines, news websites, and television often provide an interesting lead to current scholarly articles, but it is up to you to find the article or study the original source to review it for yourself. This is especially the case when members of the press seek to describe and interpret the findings of a report. They often misquote or take the original author out of context in their rephrasing. Internet blogs and other websites can provide a mixed bag of sloppy research and legitimate, trustworthy information. Information that appears on websites often does not have to undergo peer review to be checked for errors. Internet sites do not have to adhere to any rules prior to posting information. So as a researcher trying to find the best and most accurate information, beware. It is important to consider the quality of your sources both for the sake of your project's excellence and for your reputation as a careful researcher. You would not want your paper to be built on weak research, and you would not want others to think poorly of your professionalism.

How to Cite Sources

Different disciplines use different styles of citing sources. Common styles found in the journals read by sociologists are those of the ASA, American Psychological Association, and Modern Language Association. In each of these styles there are different sets of rules for formatting, but there are common requirements about the kinds of information needed. This section will help you understand and apply the ASA standards to citing information in a written assignment.

The best source for information about citing sources and formatting papers in any discipline is that discipline's style manual. Many disciplines and university departments now also have supplementary websites to accompany the style manuals, often including the latest updates for citation formats. A quick Internet search for ASA guidelines will lead you to some of the best. I have summarized here the most commonly needed pieces of information for the completion of a thesis or course paper, but I urge you to consult the official ASA style manual.

In-text Citations

Formal academic writing in the social sciences includes frequent signposts inside of parentheses, indicating the author and year of the citation being referenced. These signposts are known as in-text citations, simply meaning that in your text, these are explicit flags that tell the reader where in the reference section to find the full citation. For the ASA style of referencing, in-text citations include the author's name and year of publication.

Examples

Author's name in a reference

> More than a half million people living in Canada trace their origins to the Caribbean (Plaza 2009).

Putting the author's name in text

> Plaza (2009) reported that . . .

Multiple authors, repeated

If a large team of authors (three or more) is cited more than once, you can shorten the later citations like this, using *et al.* to replace the additional authors:

> First mention: Some have argued that unclear communication leads to disunity within the organization (Edwards, Torgerson, and Sattem 2009).
> Later mention: Edwards et al. (2009) also argued that organizations . . .

Direct Quotes

If you use a direct quote, word for word, then you use quotation marks. When you take direct quotes like this, you should provide page numbers.

> "African origin Caribbeans tend to elicit negative images in the consciousness of the dominant population whereas Indian origin Caribbeans are regarded as closer to the model minority" (Plaza 2009:38).

> Plaza (2009:39) notes that "second generation Caribbean-origin men and women in post-secondary institutions create web sites in order to give themselves a new voice to disseminate information about their Creole culture, history of migration and transnational lifestyles."

Block Quotations

Although excessive use of long quotes is frowned on in most research papers, sometimes it is important to quote a large part of a text. If you are using long quotations you need to indent the block. Block quotations should not be enclosed in quotation marks. The author, date, and page number follow the period in a block quote. Here's an example:

> Safety is hard to reward because it does not obviously contribute to the company's bottom line. Indeed, safety failures or high accident rates detract from the bottom line, but because organizations can only imperfectly measure how safe a worker is being, they only symbolically and occasionally reward a worker for appearing to remain accident-free. As a result, in practice, "safe" is a necessary but insufficient characteristic of a worker who will be punished if found wanting, but not necessarily rewarded if excellent. Indeed for the whole organization, "safe" is a necessary but insufficient characteristic for survival. (Edwards and Jabs 2010:709)

The Reference Section or Bibliography

The list of full citations may be called the bibliography or the references page. Notice when you read journal articles how different journals use different titles to describe this last section of the article.

One thing you will always see in these sections is this: The citations are listed alphabetically by the last name of the first author. Make sure your bibliography is formatted this way and does not list the citations in the order in which you discovered them or the order in which they are introduced in the paper.

Now, how do we format those full citations? It depends on the type of citation used.

Citing Books[1]

The basic form for a book entry is (1) author's last name, followed by a comma and the first name and middle initial, ending with a period; (2) year of publication followed by a period; (3) title of book, italicized, ending with a period; and (4) place of publication, followed by a colon and the name of the publisher, ending with a period. Notice how this form is achieved in the following examples.

Examples

Book with one author

Gallagher, Sally. 2003. *Evangelical Identity and Gendered Family Life*. New Brunswick, NJ: Rutgers University Press.

Book with two or more authors

Mosher, Clay and Scott Akins. 2007. *Drugs and Drug Policy: The Control of Consciousness Alteration*. Thousand Oaks, CA: Sage Publications.

Notice here that the first author's name is inverted so one can alphabetize and the second author's name is not inverted. Otherwise, it's the same as the previous example.

Chapter in an edited book with different chapters by different authors

Inderbitzin, Michelle. 2007. "The Impact of Gender on Juvenile Justice Decisions." Pp. 782–91 in *It's a Crime: Women and Justice*, 4th ed., edited by R. Muraskin. Upper Saddle River, NJ: Pearson Prentice Hall.

Notice that here I give credit to the author of the chapter and also to the editor who assembled the book. By having the word *in* as part of the citation I demonstrate that this is a chapter in a book. Otherwise, you can see familiar elements in all of the above citations.

Citing Journal Articles[1]

The basic reference format for a journal article is (1) author's last name, followed by a comma and the first name, ending with a period; (2) year of publication followed by a period; (3) title of article in quotations and ending with a period inside the closing quotation mark; (4) title of journal in italics; (5) volume number; (6) issue number if available, enclosed in parentheses, followed by a colon; and (7) elided page numbers and period.

Examples

Article with one author

Plaza, Dwaine. 2006. "An Examination of the Transnational Remittance Practices of Jamaican Canadian Families." *Global Development Studies* 4(3/4):217–50.

Article with more than one author

Hammer, Roger B., Volker C. Radeloff, Jeremy S. Fried, and Susan I. Stewart. 2007. "Wildland-urban Interface Housing Growth during the 1990s in California, Oregon, and Washington." *International Journal of Wildland Fire* 16(3):255–65.

Notice that the first author's name is inverted so we can list the references alphabetically and the second author's name is not inverted, but otherwise this is the same as the one-author form for an article.

Article within an article

As described early in this chapter, sometimes (rarely) we mention someone's research as it was presented in another person's research. Recall the Smith and Wilson example early in this chapter.

Wilson, John. 1995. "The Impact of Education on Support for Gun Control." *Journal of Firearms* 1:100–20, in Miguel Smith. 2010. "Variables That Impact Support for Gun Control." *Journal of Policy Studies* 10:202–9.

This approach is generally discouraged for the reasons outlined above, but when it is unavoidable, this citation format at least gives both authors credit for their work.

Citing Newspaper and Magazine Articles in Print[1]

The basic format for a newspaper or magazine entry is (1) author's last name, followed by a comma and the first name and middle initial, ending with a period; (2) year of publication followed by a period; (3) title of article in quotations, ending with a period inside the closing quotation mark; (4) name of newspaper/magazine in italics, followed by a comma; (5) date of publication followed by a comma; and (6) page numbers of article within the publication, ending with a period.

Examples

Magazine

Padgett, Tim. 2010. "Who's to Blame for Suspending Haitian Medevac Flights?" *Time Magazine,* January 3, pp. 15–17.

Newspaper

Bishop, Greg. 2009. "Taking Vows in a League Blindsided by Divorce." *New York Times,* August 9, SP1.

Citing Government Documents

Government documents come in many forms—reports, papers, compendia, websites, data tables, and so forth. So it is impossible to set up a singular rule to describe them all. Here are some examples to learn from, but make sure you ask your professor or supervisor or consult the ASA style guide.

Examples

Frey, William. 2008. "A Compass for Understanding and Using American Community Survey Data." Washington, DC: U.S. Department of Commerce, Economics and Statistics Administration, U.S. Census Bureau.

U.S. Census: U.S. Bureau of the Census. 2010. "The 2010 Census and the American Community Survey: America Is Changing, and So Is the Census." *Characteristics of Population.* Vol. 1. Washington, DC: U.S. Government Printing Office. Retrieved May 18, 2011 (http://permanent.access.gpo.gov/LPS117569/LPS117569/2010.census.gov/partners/pdf/2010_acs_dropin.pdf).

U.S. Department of Justice, Office of Justice Programs. 2005. *U.S. Department of Justice Recommended AMBER Alert Criteria.* Washington, DC: Office of Juvenile Justice and Delinquency Prevention.

Citing Research Presented at Meetings and Symposia[1]

Some research is presented at conferences but is never published. However, an author may hand out copies of his or her paper, and you may obtain one. Or an author may e-mail you a copy. For contributions at these gatherings of academic researchers, use the following: (1) presenter's last name, followed by a comma and the first name and middle initial, ending with a period; (2) year of presentation followed by a period; (3) title of paper or presentation, italicized, ending with a period; (4) "Paper presented at the meeting of [organization name]," followed by a comma; (5) month and date of presentation, with comma; and (6) location, ending with a period.

Edwards, Mark E. 1999. *Accelerated Growth in Employment of Preschoolers' Mothers: Real & Perceived Need.* Paper presented at the meeting of the Pacific Sociological Association, April 5, Portland, OR.

Citing Dissertations and Theses

Doctoral dissertations and master's theses can be very important sources for sociological research papers. These sources can provide a good review of the literature and give you a good place from which to start writing your paper. Sometimes these can be found in your school library or, increasingly, in university data archives.

Examples

Edwards, Mark. 1997. "Toward Explaining Accelerated Rates of Employment among American Mothers of Preschoolers: 1965–1988." PhD dissertation, Department of Sociology, University of Washington. Retrieved from Dissertation Abstracts Online, AAG9736263.

Porter, Suzanne. 2010. "The Dynamics of Work, Poverty and Business Cycles: An Analysis of Oregon Households Receiving Food Assistance." Retrieved January 3, 2011, from Oregon State University Scholars Archive (http://hdl.handle.net/1957/14921).

Citing Technical and Research Reports

Technical and research reports, like journal articles, usually cover original research but may or may not be peer reviewed. These reports can be very important in sociology papers, supplementing other findings in peer-reviewed journals. References for most technical reports will be formatted similarly to book or article references but should include other identifying numbers provided by the issuing organizations.

Example

Grussing, Jay and Mark Edwards. 2006. "Non-metropolitan Hunger and Food Insecurity in the Northwest." Working Paper No. RSP 06–02, Oregon State University Rural Studies Program. Retrieved February 15, 2011 (http://ruralstudies.oregonstate.edu/working-paper-series).

Citing Articles, Newspapers, and Other Materials Retrieved in Electronic Format

Items that have been found on the Internet require some additional information. Usually this additional information includes a web address or the name of a particular search engine or database as well as information about when the document was last retrieved by the researcher. This last piece of information is needed because electronic sources tend to move around, be made more or less available, get updated, and sometimes disappear.

Examples

Academic article (usually in print) in electronic format

Many journals that actually print hard copies and mail them out also have electronic subscriptions available through academic libraries. With official permission (usually to registered students), you can access those journals online without having to go to the library. The electronic databases of these files permit you to find and read them. Some instructors require that if you access a journal article via a commercial database, you include the name of that database (e.g., EBSCO-host, LexisNexis). When instructing my students, I do not require this. But if your instructor requires it, it is best to do what you're asked! Notice that the word *Retrieved* plus the date that you accessed the paper, as well as the name of the database, are the additions here.

Brunson, Rod K. and Jody Miller. 2006. "Gender, Race, and Urban Policing: The Experience of African American Youths." *Gender & Society* 20(4):531–52. Retrieved December 15, 2010 (http://gas.sagepub.com).

Sundquist, Christian. 2009. "Defining Race: On Race Theory and Numbers." *Albany Law Review* 72:1. Retrieved November 29, 2010 (http://www.lexisnexis .com/). LexisNexis Academic Universe, Law Reviews.

Web-based journal

Some journals are completely online, with no hard copy sitting on a library shelf.

Plaza, Dwaine and Kathleen Stanley. 2002. "Camaraderie and Hierarchy in College Football: A Content Analysis of Team Photographs." *Sociology of Sport On Line* 5(2). Special Issue, November/December. Retrieved November 28, 2010 (http:// physed.otago.ac.nz/ sosol/v5i2/v5i2.html).

Newspaper in electronic format

This example and the subsequent examples of references for web-based items require the website and the date retrieved.

Drogin, Bob and April Choi. 2010. "Mixed Portraits of Oregon Terrorism Suspect." *Los Angeles Times,* November 28. Retrieved November 29, 2010 (http://www.latimes.com/news/nationworld/nation/la-na-oregon-bomb -plot-20101129,0,2189239.story).

Report posted on a website

Scelza, Janene and Roberta Spalter-Roth. 2010. "The Gap in Faculty Pay between Private and Public Institutions: Smaller in Sociology Than in Other Social Sciences." Washington, DC: American Sociological Association. Retrieved November 15, 2010 (http://www.asanet.org/research/facsaldatabrief.pdf).

Admittedly, producing the reference section or bibliography of a paper can be tedious. There is software available that helps with this task, and this is especially useful for very long projects. For most term papers, those programs provide more power than you need. Some careful note taking as you develop your paper and maybe the help of a friend who can watch for inconsistencies in your formatting will help you efficiently produce a clean, professional-looking list of references that demonstrates how thorough you have been in your work. When in doubt, consult an official style manual for your discipline, and ask the community of writers of which you are a part.

Note

1. The descriptions that articulate the elements of each kind of entry are taken from Salinas, Romelia. 2010. *ASA Format.* Retrieved February 18, 2011 (http:// www.calstatela.edu/library/bi/rsalina/asa.styleguide09262007.html).

Section II

Writing Quantitative Papers

Chapter Six

Quantitative Papers

The Introduction

The introduction to an academic paper is the place where you try to hook the reader with an answer to the question, "Why should I care?" This is also the place where you identify the main issue that you will address in your paper. Introductions are challenging to do well because you must strike a balance between promoting your topic enough to convince the reader that this is really worth reading and avoiding overstating your case.

If you were indeed a prosecuting attorney, you might want to appeal immediately to questions of justice, to use strong emotional language, or to direct the jury's attention to those "fat cats sitting smugly there in their expensive suits." However, in formal research writing we do not use such grand gestures to attract readers' attention. Indeed it is a common rookie mistake in writing introductions to such papers to try to persuade readers by appealing boldly and/or solely to moral arguments, excessively relying on the heart-tug element of the issue. Often the drama or tragedy of the issue is overstated. For example, consider this introduction to a paper about mothers, fathers, new children, and work:

> Sixty-four percent of mothers of preschoolers are in the labor force (U.S. Census 2010). Therefore, millions of families across the country struggle every day with the conflict of work and family and agonize over whether to let other people raise their children or stay home and perhaps damage their own careers. Why do they do it?

Any or all of these claims may be true, but there are lots of loaded, and likely overstated, claims that threaten the author's credibility. Indeed, the

64 percent statistic is correct, but whether this translates into millions of people struggling with a conflict is not obvious, and it may be overdramatizing the issue to say that people are agonizing over letting others raise their kids. Even if it is true, this style of writing comes across as either preachy, grandiose, or tabloid-esque. Finally, while a rhetorical question can be useful in some cases, here it strikes a tone of shocked criticism, as if the author were saying, "Holy cow! What were they thinking?"

The writer could communicate the same issues in a more even-handed way that would invite readers who agree or disagree to continue reading further. For example, try this as a possible improvement:

> Sixty-four percent of mothers of preschoolers are in the labor force (U.S. Census 2010). While many families appear to be juggling the work-family conflict adequately, others claim to feel guilty about leaving their children in the care of other adults and perhaps missing out on important events in their young children's lives. Meanwhile, the potential setbacks in their careers make it difficult for young parents to consider taking time out of the labor force.

This revision is not perfect, but it avoids some of the inflammatory speculation about agonizing and avoids making it sound like the author is accusing people of letting others raise their kids. Even if that's what the author thinks, she or he would need to carefully consider the audience and ask whether the tone being struck will inspire or turn off readers.

There may be times when the nature of your writing should provoke response. But make sure that you are choosing this for some other effect than just to get people to read further. If you alienate your readers in the first paragraphs, they may just toss the paper (or be put in a foul mood when grading it). So beware of excessively provocative language and tone, and watch out for overstatement that might damage your credibility.

Let's add to this revision some material about what the paper will do:

> Sixty-four percent of mothers of preschoolers are in the labor force (U.S. Census 2010). While many families appear to be juggling the work-family conflict adequately, others claim to feel guilty about leaving their children in the care of other adults and perhaps missing out on important events in their young children's lives. Meanwhile, the potential setbacks in their careers make it difficult for young parents to consider taking time out of the labor force.
>
> This paper identifies the characteristics of young mothers and fathers that are associated with full- and part-time employment while the first child is still an infant. Unlike earlier studies that rely on cross-sectional data, this analysis follows the early life histories of young families to locate how not only demographic characteristics but also the timing and order of events influence the likelihood, for new mothers and fathers, of returning quickly to paid work.

Notice that this revision introduces the topic, sets some of the context for why we would care, and then goes on to state briefly what the analysis is about.

You'll notice that unlike a mystery novel, a quantitative research paper uses the introduction to give away the plot right away. It may or may not tell the ending and conclusion, but it makes clear to the reader where this is going. In this example, we use the present tense in the last paragraph. Others prefer a past or future tense. Your professor or editor can guide you about what he or she wishes to see.

In my opinion, this sample introduction needs to be further elaborated, including material about other interested parties (employers) and about how this paper relates to the existing body of knowledge about this topic. Consider the next revision. The goal has been to show that the topic is important, interesting, and newsworthy but without asserting these things in a heavy-handed way. I've italicized the new material I added about other parties and previous research.

> Sixty-four percent of mothers of preschoolers are in the labor force (U.S. Census 2010). *More than 90 percent of fathers of preschoolers are in the labor force (U.S. Census 2000).* While many families appear to be juggling the work-family conflict adequately, others claim to feel guilty about leaving their children in the care of other adults and perhaps missing out on important events in the young children's lives. Meanwhile, the potential setbacks in their careers make it difficult for young parents to consider taking time out of the labor force. *Employers are also concerned about this issue as the state continues to pass and consider new laws providing family leave and as they seek to retain skilled workers.*
>
> *Most previous research has emphasized the human capital arguments for new mothers' rapid return to work. However, little effort has been made to understand how the order of events such as parents' educational attainment, cohabitation, marriage, first job, promotions, and the like are related to the decision of mothers to remain in the labor force. And no research has explored how these characteristics influence the likelihood that new fathers will take time off to be with their new children.*
>
> This paper identifies the characteristics of families and young mothers and fathers that are associated with full- and part-time employment while the first child is still an infant. Unlike earlier studies that rely on cross-sectional data, this analysis follows the early life histories of young families to locate how not only demographic characteristics but also the timing and order of events influence the likelihood, for new mothers and fathers, of returning quickly to paid work.

The second paragraph briefly states how this paper is an improvement on previous work. Your paper may be an improvement on or a replication of previous work—either is fine. Just specify what you think. The last paragraph also emphasizes how this paper is an improvement and lets the reader know what to expect in the methodology. It also makes clear what the research question really is, namely, "What personal characteristics are correlated with full- and part-time work for new fathers and new mothers, and how does the order of prior events influence the likelihood of that work?"

To summarize, a good introduction accomplishes the task of introducing the topic and convincing the reader to believe that this is worth learning about without appealing to overdramatizing the importance of the project. And it points out why this research is worth doing and makes it clear that you did not do it just because it was there. Try reading the introductions to several research journal articles to see how they do it. You will find different strategies and indeed a wide variety in quality of introductions, showing just how challenging it is to write a good introduction.

Finally, I offer some precious advice I once received from a graduate mentor (Becker 2007). Write the introduction after the paper is done. Papers shift directions as they are being produced, and too much time is often wasted writing the perfect introduction to a paper that never actually turns out to be what was introduced. It is not uncommon to find unpublished manuscripts by professors and final papers turned in by students in which the introduction and the paper do not really match because the author never checked back after writing the paper to see if the introduction still introduces the paper.

Reference

Becker, Howard S. 2007. *Writing for Social Scientists.* Chicago: University of Chicago Press.

Chapter Seven

Quantitative Papers

The Literature Review

Different judges may disagree over exactly how much instruction they think jurors need to be able to render a verdict informed by the law. But smart attorneys, aware of whatever rules judges set, make strategic decisions about how much information and background to present to jurors and in which order to present it. From the vast array of previous legal precedent and from the existing applicable laws, the attorney must decide how to explain it to jurors so that they will know how to interpret the evidence to be presented.

There is a parallel challenge you face as a writer of sociology. Sociology instructors and others who request sociological writing (such as philanthropies, government agencies, and your supervisor) often disagree over what precisely one should include when writing a literature review. This disagreement can pose a problem for you as you write. So first, be aware of who is setting the rules and what he or she expects. If you have to keep it short, that expectation will suggest a different strategy than if you are asked to be expansive and inclusive. If you are writing for a well-informed audience then you will include different things than if you are writing for a general audience unfamiliar with your topic. Most of the decisions you will make revolve around these three things: What should be included? How long should it be? and How should it be organized?

Thankfully, there are some common expectations among many of the authorities we write for. At the minimum, a literature review is the place

Author's Note: Thanks to Rebecca Warner for help in creating an earlier draft of the chapter.

where the author situates the research paper in the existing research literature. That is, this is the part of the paper where you help the reader see how what you are doing is related to the projects that other people have completed.

In chapter 4, which addresses borrowing well from the literature, I used the analogy of summarizing conversations between researchers. As a writer, you have a lot of influence over how to recount those conversations. You can be strategic and creative in how you arrange that summary of the research. As you read different published research papers, you will see that the organization of materials in a quantitative paper is likely to be around the different variables that are measured and analyzed. This chapter focuses on a very typical structure of a quantitative literature review wherein the existing conversations in the literature are arranged in a strategic way.

A Flexible Prescription for a Literature Review

Many quantitative social science literature reviews in academic articles and books, as well as in grant proposals, graduate theses, and government agency reports, tend to be organized first around the dependent variable; then they proceed to discussions of how other variables may affect that dependent variable. This makes sense because we often write about cause-and-effect questions such as, "How much does education influence attitudes about abortion?" or "How big an impact does this arts program have on children's engagement in school?" That is, the effect (attitudes, engagement, etc.) is the important outcome we are interested in (hence, we discuss it first), and we are seeing which variables influence that important outcome (hence, we subsequently turn attention to those other variables). Take a look at the general outline for organizing a literature review (see p. 49), and then consider flexing it a bit as you need to. With the addition of other independent variables, this outline would expand further.

The opening paragraphs of a literature review (item I in the Outline for Literature Review box) usually accomplish at least a couple tasks. First, they help us conceptualize the dependent variable, recounting how others have conceptualized it, argued about its definition, or otherwise puzzled over it. Second, they help the reader to understand the current status of the dependent variable out in the population.

If you were writing about (rather than prosecuting the perpetrator of) age discrimination, you would use the opening paragraphs of the literature review to help readers understand what age discrimination is, how people have defined it, and who does the defining (e.g., How have social scientists,

Outline for Literature Review

I. Discussion of dependent variable (conceptualization, parameters, trends)

II. Influence of first independent variable on dependent variable
   a. Theoretical rationale, predicted relationship between variables, and causal mechanism by which independent variable affects dependent variable

III. Influence of second independent variable on dependent variable
   a. Theoretical rationale, predicted relationship between variables, and causal mechanism by which independent variable affects dependent variable

IV. Relationship of independent variables and rationale for controlling one or more
   a. Theoretical rationale, predicted relationships between variables, causal mechanism, and logical rationale for controlling for one

rather than lawyers or politicians, defined it? How old must a person be for one to regard discrimination as a case of age discrimination?). Then you could report what is known about the level of age discrimination in society or the number of lawsuits or legislative actions related to age discrimination and how that might be changing over time. By the end of the opening couple paragraphs the reader should be able to say, "I understand which dependent variable will be studied, I understand some of the complexities in defining this variable (and what the author thinks about that), and I have some sense of how the attributes of the dependent variable are distributed out in society."

Typically, quantitative sociological papers seek to explain why a dependent variable varies. That is, why is there more of it in one group or place than another, or why does it covary with other variables? Often sociological theories help us to anticipate what those other important variables are, how they will vary with our dependent variable, and why. So theory usually begins to appear in the next paragraphs, justifying both the exploration of how a particular independent variable will have an impact and the means by which we think the independent variable influences the dependent variable.

In the case of age discrimination, let's say we are interested in which groups are more or less likely to experience age discrimination. Perhaps we know something about Marxist feminist theory that tells us that in a market economy, people are valued for what they can produce and that women are often regarded as problematic workers because they more often end up having to care for not only kids but aging parents. As a result, we might predict that women would experience age discrimination more often than men because employers might be less accommodating for employees who have extensive family obligations (such as elder care), and thus women in the latter years of their careers would more often experience age discrimination. Notice that the theory has helped us to select an independent variable (gender), to make a prediction (women more than men), and to provide a theoretically informed causal mechanism (employer bias combined with women's culturally prescribed family care roles).

So in this next block of paragraphs (item II in the Outline for Literature Review box) we would make the case that, rooted in Marxist feminist theories about workplace relations, we would expect women more than men to experience age discrimination. In these paragraphs we would cite these arguments from theorists and present to the reader existing research findings that give us reason to anticipate in our study that this is what we will find. By this point, the reader should be able to say, "The author expects to find that women more than men experience age discrimination, and here's why." (See chapter 4, "Borrowing Well from the Literature," for more help on this topic.)

Next, we turn our attention to additional variables that also are likely to be important (item III in the Outline for Literature Review box). Presumably theory should help us select those variables. That's not always the case. Sometimes our interest in other variables is not inspired by a theory, but their importance or our interest needs to be logically defended. Sometimes we wish to control for a variable to determine whether a correlation that exists in the literature is spurious or somehow suspect. For example, if others have found gender differences in age discrimination, perhaps you think it's not because of gender bias but because of the kinds of jobs women versus men hold. Let's say that our reading from Marx also tells us that the kinds of jobs people have make a big difference in terms of their control over the flexibility of their jobs and also in terms of how employers will think about these employees. That is, we might predict that managers and professionals will have a better chance of avoiding or fighting off age discrimination than people in jobs that are physically demanding or jobs for which it would be relatively easy to find a replacement. That line of argument would help us hypothesize that professionals and managers are less likely than other workers to experience age discrimination. As in the previous paragraphs, our goal here is to justify why we are looking at type of job as an influence on likelihood of experiencing age discrimination, to hypothesize an anticipated pattern, and to provide some rationale for why this is what we expect.

The remaining couple paragraphs (in this simple example of a paper with only three variables) help the reader understand how gender and type of job are related to each other and what we expect to learn by controlling for type of job. Here we could again draw on theory to help us explain why women are more concentrated in some jobs than in others, but we can also cite existing research that reports statistics on occupational sex segregation. By controlling for type of job, we may anticipate that the gender gap in age discrimination will be smaller in both blue-collar and managerial/professional jobs, showing that some of the apparent gender effect is due to the jobs that women versus men hold. This is where things get really interesting analytically, and it is the job of the sociological writer to explain why we control for this variable and what we should expect.

By the end of the literature review, readers should be able to recount the main independent variables, their predicted effect on the dependent variable, and the reasons for considering more than one independent variable. In other words, when they turn to the next section of the paper (Data and Methods) they should have a sense of what to expect.

Using the Literature to Develop a Causal Model

Once we have some idea as to the order of the paragraphs in the literature review, we can start to arrange the existing research materials in a way that conveys the relationship between variables. One technique for sorting out what we have found in the research literature is to first make a list of the findings (empirical observations and patterns) and of theoretical claims found in the relevant citations that you have located. Let's continue with our earlier example about age discrimination. Below is the causal model that we just articulated, where experiencing age discrimination is believed to be affected by gender and by the jobs that workers have.

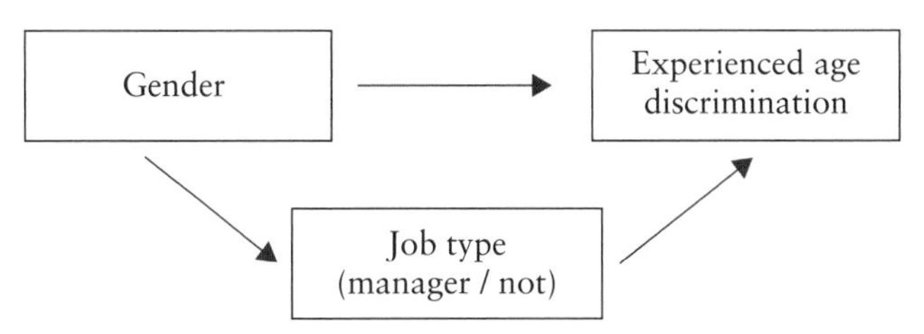

The picture merely communicates to us that (1) one's gender and job each affect the chances of experiencing age discrimination and (2) because gender and job are related, it's possible that the effect of gender is partially explained by the job one has (i.e., perhaps women are less likely to be managers, in a position to enjoy a less rigid job schedule).

Now let's consider these statements that one might be able to write down after scanning the research literature on related topics. (Note: These are made-up citations for illustrative purposes only.)

Juarez (2001): The odds of being a manager are twice as high for a 50-year-old man than for a 50-year-old woman.

Krzynski (2005): Qualitative interviews of California assembly-line workers and managers reveal that assembly-line workers have greater struggles than managers at juggling work-family conflicts.

Tongo (2007): Age discrimination lawsuits in a three-year period in Ohio reveal that two thirds of the plaintiffs were workers with very little managerial authority.

Schnell (2008): In a multistate region in the Northeast, secretaries were more likely than their female supervisors to describe ways that their organizations made it difficult for them to take care of family obligations.

Habab (2006): Employer surveys indicate that there is a bias toward younger workers, with employers believing that younger workers are more energetic, have fewer family obligations, and cost less.

Lighthouse (1999): In private interviews, employers admitted that they found it difficult to accommodate workers who had repeated work-family difficulties and that elder care issues were as common among their workers as were child care issues.

Oleander (2008): In a nationally representative sample of American workers, female employees were twice as likely as male employees to be able to recount stories of how employers had made it difficult for them to get time off to deal with family demands.

Palanka (1997): Throughout the life course, women repeatedly are socialized toward caregiving roles and are given fewer opportunities to move up into management.

Grey (2002): Age discrimination is harder to prove in court than other kinds of employment discrimination cases.

Sanchez (2000): Marxist feminist theorists argue that women are regarded by employers as problematic because of their usual family responsibilities.

When we have a list of such findings, we then can consider how these items can be assembled, keeping in mind (1) the loose structure suggested earlier and (2) the goal of recounting a conversation that one can imagine in the literature.

For visual learners, it might be useful to write down the different citations (author and year) on the different arrows in the diagram that are explained by those authors. For example, the Juarez study illustrates the arrow that links gender and job. Other sources may not illustrate an arrow but rather might help describe a variable or concept, such as the Habab citation that emphasizes employers' mindsets. Imagine those kinds of citations going on top of the boxes. By sorting out resources this way, you can strategize how you'll use the array of resources located in a literature search.

See how in the following draft I try to assemble these items, and notice how I have sorted out the items shown into the locations that illustrate each relationship (arrow) or variable (box):

> While many characteristics of employees are rewarded and punished in the workplace, age can be the most confusing for employers. Habab (2006) reports that employers generally have a bias toward younger workers because of their perceived skills, minimal family obligations, and lower salaries. Whether employers specifically and actively discriminate against older workers is not always clear. Grey (2002) points out that compared to other kinds of employment discrimination, age discrimination is particularly difficult to prove. [Notice that I use Grey and Habab to conceptualize the dependent variable, age discrimination.]
>
> Because employers value having workers with few family obligations, there is good reason to anticipate that among older workers women will experience age discrimination more often than men (Sanchez 2000). Ongoing processes of gender socialization and persistent family obligation patterns make women more likely than men to have ongoing work-family obligations because of the problem of elder care (Palanka 1997). Women in general report higher rates of inflexible treatment by employers (Oleander 2008). Evidence from employers supports this expectation as well. Lighthouse (1999) reports that employers admit to having difficulty being flexible with workers and that elder care issues are as common now as child care issues. Thus, it is likely that as older working women embrace the culturally prescribed elder care responsibility, they will become more difficult employees to keep, and employers will more likely find ways to either release them or avoid hiring them. These processes are likely to be less pronounced for men. [Observe how I use four of the sources to focus on the hypothesis that gender would affect age discrimination.]
>
> However, the gender difference in age discrimination may be partially explained by the kinds of jobs that men and women have. Job authority can have a significant impact on the chances of experiencing work-family conflict. Krzynski (2005) and Tongo (2007) find that work-family struggles and age discrimination both seem to be associated with having less workplace authority. Schnell (2008) found that among female employees, secretaries more often than managers identified ways that they experienced the inflexibility of their employers. Therefore, there is good reason to believe that the direct effect of workplace authority would influence older employees' experiences of workplace demands and hence of age discrimination by employers. [In this paragraph, I focus on three other sources that link job and age discrimination.]

> Because older women are less likely than men to be in positions of managerial authority, they may be more vulnerable to age discrimination. Juarez (2001) computes that for older men the odds of being a manager are twice those for older women. The processes by which this pattern emerges are well known. . . . [Here I start focusing on the other causal arrow—gender and job.]

Notice that in assembling the literature review this way, I have created some conversations, such as in the reference to Krzynski, Tongo, and Schnell, all of whom had insights that illuminated one idea. Or consider the second paragraph wherein you might imagine getting Palanka, Oleander, and Lighthouse to talk and see just how consistent their findings are.

In this example perhaps you can see how an author can capture a sense of the ongoing, sometimes disorganized set of conversations out in the literature and do so in a way that provides the rationale for a researchable question. By doing this, one avoids merely quoting the other authors' words but reports efficiently the substance of their findings and conclusions.

As you read articles that will be referenced in your research, analyze a few literature reviews to see how they set it up. You will observe that there is a strategy to how skilled writers set up their literature reviews so the reader can anticipate what will be included in the subsequent analysis.

Chapter Eight

Quantitative Papers

The Data and Methods Section

Some people doubt that sociology is in fact scientific. The Data and Methods section of one's research paper makes it clear that this discipline emphasizes some of the most central characteristics of an empirical science—carefully described, repeatable methods that allow other researchers to check one's work and replicate one's findings. This is an obvious place that my courtroom analogy breaks down. Lawyers are not likely to say to the jury, "And to get these data, we wrote three formal letters to the employer before the company finally provided us what we wanted." They might occasionally talk about how forthcoming someone was in providing evidence, but they are not required to. Sociologists are required to make clear how they obtained their data, make clear how they measured the things they are discussing, and help the reader understand the wisdom in doing it the way it was done.

The Data and Methods sections of most research papers focus first on the data (the sample), then the measures, and then the actual analyses conducted.

Data (the Sample)

The initial emphasis on data makes it clear why sometimes this section of your paper is titled Data and Methods. Sociology is empirical, and to be empirical, you need data. Hence, we begin with the data used. This, by the way, is true whether we are collecting quantitative or qualitative data. Our focus in this chapter remains on quantitative research.

Sociologists do not often have access to the entire population of interest. Instead, they have the time and opportunity to gather data only from a subset of that population, that is, a sample. So in the Data and Methods section, you tell the reader how you located that sample and what were the characteristics of the sample (Was it obtained as a probability sample? Nonprobability? etc.). The discerning reader will then expect you to answer these questions: To whom will you be able to generalize your results? Are you generalizing to all American adults? To adults of a certain age? To employed adults? To all households with a telephone? To students who happen to pass the student union on a Monday at midday?

Of central concern is whether the sample used is appropriate for the research. While ideally you may wish to go get your own data, very often you will be using data collected by someone else. When this is the case, you provide the reader with the name of the data set and how the sample was drawn. You may want to explain why this data set was most appropriate for answering the question that you are posing. That is, how do these data help you meet your research goals? Consider the following examples:

> The 2008 Current Population Survey March files are especially useful for answering the central research questions because the large sample sizes permit comparisons between occupational subgroups. Other smaller data sets provide too few cases to compare specific occupations such as nurses versus secretaries versus factory workers. . . .

> The National Survey of Baby Boomers provides the necessary information covering each working adult's work conditions, the work-family policies of the organizations they work for, and their opinions about how accommodating their employers have been. These data are not included in other readily available data sets.

In addition, you may have drawn your own subsample from the secondary data. You may be studying women and men between the ages of 45 and 70 or adolescents age 12 to 17. If you are taking a subsample, be sure to articulate why you are selecting this subsample. How does it relate to your research questions? The following is an example:

> Because I am interested in age discrimination among working-age people, I limit the age range to employees who are 35 to 70 years old. Blank (2002) and Wilson (2003) both focused on this same age group, so it is possible to compare my findings with theirs. By excluding people older than age 70, I preclude the study of many retirees.
>
> Self-employed men and women are eliminated from this sample since the focus here is on men and women potentially facing age discrimination by employers.

Notice that sometimes in the Data and Methods section, we include other citations from the literature because other social scientists have made methodological decisions that inform our research.

It is important to avoid explaining and defending our samples with weak arguments. Here are some weak ones:

> The Panel Study of Income Dynamics was used because it has many variables that might be of interest and because it is commonly accepted by social scientists as a trustworthy source of information.

> I focus attention on Hispanic men because the data do not include enough Asian men for me to directly answer the questions about education and social mobility.

Indeed, the comments about the Panel Study of Income Dynamics data are true: There are lots of variables, and the data set is widely accepted, but these are not good enough reasons for using it. The reasons must derive from the research you are doing, focusing on why these data are good for answering your question. Meanwhile, you do well to avoid pointing out your good but unfulfilled intention ("I really wanted to study Asian men") and instead make it clear that the group you are studying is theoretically and substantively defensible as a group to study. If, in this case, it became clear that only Hispanic men would be available in the data, the whole paper should take this into account from the very beginning (and you ought not bring up your original intention here as an afterthought, excuse, or complaint about what might have been).

Measures

If you are using survey data, this section should include the questions asked of your sample of respondents, along with how their responses were categorized. The major concerns that a reader will have, when learning about your measures, are these: How valid and reliable is this measure for capturing the different variables included in the study, and how precise is the measure?

For example, let's say you are interested in how much education respondents have. So you use the question, "What is the highest level of education you have received?" The reader will want to know if asking people this question will obtain accurate, truthful answers that are more or less prone to misreporting. This one is pretty straightforward, but if you were measuring whether some experienced age discrimination, you would need to explain why asking a person if he or she was mistreated may or may not accurately tell you who was discriminated against. As you read research papers, watch how authors defend the reliability and validity of their measures and how they sometimes admit to weaknesses in the measures.

With regard to precision, a common concern is whether you have taken data that were collected using a more precise measure and converted them into a less precise measure. That is, have you collapsed the data into fewer categories (perhaps to simplify the analysis)? For example, you could collapse highest grade completed into less precise educational categories so that responses include (1) less than high school, (2) high school diploma, (3) some college, (4) college degree, (5) some graduate school, and (6) graduate-level degree. You certainly will have reasons for why you did this, and why you did it in this particular way, and you should articulate those reasons for the reader. For example, you could have recoded education as less than college degree versus college degree. Recoding categories depends on a number of things. One is sample variability (whether the data are skewed); another is theoretical relevance (income does not respond linearly to highest grade attained, but it probably does go up steadily across different educational degrees); and another may be for statistical procedure requirements (logistic regression versus ordinary least squares regression versus simple tables with percentages).

Here's an example:

> Because of my effort to identify middle-class respondents by their educational attainment, I collapsed those with less than a bachelor's degree into one category and all respondents with at least a bachelor's degree into the second category. College completion is widely accepted as an indicator of middle-class identity.

Notice that I make a vague, bold claim that there is good reason to use college completion as a cutoff point (although reviewers might think I should further defend that point).

Here's another example:

> Because there are so few people with graduate degrees among the blue-collar occupational groups, graduate degree recipients were clustered with all other college graduates.

Here I articulate pragmatic reasons for collapsing categories, namely, the data will not allow me to examine blue-collar workers with graduate degrees because there are so few in the data set. So I explain that I want to keep them in the study and so lump them in with the other blue-collar college graduates.

Typically, the discussion of the variables introduces the variables in the same order in which they were introduced in the literature review, beginning with the dependent variable and then each of the independent variables.

Strategy for Analysis

Finally, this part of the Data and Methods section explains how your data are analyzed. Are you calculating percentages and means and comparing them? If so, how do you determine if the differences are statistically significant (e.g., which statistical tests are you using?)? If you are using more than a couple variables, are you using multivariate tables or regression analysis? These are the kinds of things the reader will be looking for at this point in the report.

You need not explain your statistic at length unless it is unique or unfamiliar. In general, assume the reader is a social scientist who knows basic statistics or that the reader could, without much effort, look up online the details of the statistical technique. In some circumstances, you may conclude that your audience needs some extra help, but that's unusual.

Sometimes a study examines how different coding of variables results in different findings. For example, you might be checking to see if age discrimination appears to be more or less evident between men and women when we measure it using age 50 as the cutoff versus age 55. If so, then you will want to explain that you will run the analysis using the one coding and then run it using the other.

Sometimes a study is simply controlling for a variable to see if there is an impact on the initially observed relationship. In this case, the author would briefly outline that process. For example,

> I first establish the relationship between gender and age discrimination and then control for the occupation of the worker to see if the relationship diminishes as predicted.

This part of the Data and Methods section is usually not as long as the other two, in part because the author will have implied this method by a well-written literature review.

Incidentally, you do not need to tell the reader the name of your software (e.g., Excel, PAWS/SPSS, SAS) unless the statistical package is unique and handles the data in a special way. For the relatively new social science researcher, there is a strong temptation to highlight the recent success of having mastered the computer software. However, the seasoned social scientist reading your paper will assume that you used appropriate computer software to conduct the analysis.

Chapter Nine

Quantitative Papers

Presenting Results

If you have ever been on a guided tour of a museum, a theater, or some other tourist attraction, you know that the guide can make or break your experience. The excessively detail-oriented guide will put you to sleep, the ill-equipped comic guide will anger or embarrass you, and the overly casual guide will leave you puzzled and frustrated. But a guide who manages to walk you through the attraction, highlighting the most important and interesting features and weaving a coherent story that links together the parts of the attraction, is the guide who will give you the best possible tour. In the same way, the Results section of your research paper is the attraction that readers have come to see. Your task as guide is to walk them through your analysis in a coherent, deft, and efficient manner. This chapter will alert you to some of the issues involved in achieving this level of sophistication in writing about quantitative data.

The Goal

The goal of the Results section is to clearly, forcefully, and modestly communicate to your readers what the data reveal. Notice that this goal follows from the goals of your literature review (to establish what needs to be studied, why it is important, and which results are predicted) and the goals of your Data and Methods section (to describe the data set and how it was analyzed). Meanwhile, communicating what the data analysis reveals sets you up for the goal of your Discussion and Conclusion sections (to sum up the larger theoretical points raised and addressed by your analysis).

Writing about quantitative results is unlike the writing that we are usually taught in school. As opposed to writing an essay with thesis statements and supporting points, you here find yourself alerting the reader to things that you now consider obvious and that you have sought to make obvious in your tables and figures. Yet you now must point these things out to the readers. Additionally, you must allow for readers to verify your claims while at the same time not beating them over the head with endless details that they could get themselves by reading your tables and figures. So there is a delicate balancing of your efforts to point out the important points while at the same time respect your readers' ability to consider the facts for themselves. This is difficult to do well, and like many things, it only improves with practice.

An Extended Example: Redundancy and Voice

Let's follow an example derived from a student paper researching American men's incomes. In Table 9.1 below, we find the descriptive statistics for incomes of older and younger men. The author wants to comment on the difference between the two groups of men.

Table 9.1 Income for American Men

	Mean	*Median*	*25th Percentile*	*75th Percentile*	*n*
Young men (18–40) Income	36,531	30,000	19,700	43,000	10,767
Older men (41–65) Income	48,082	37,000	24,000	55,000	10,887

Here is the author's first draft describing the data:

> My analysis shows that the mean earnings for the younger men is $36,531, and for older men it is $48,082. This is a difference of $11,551. There were 10,767 younger men and 10,887 older men in the subsamples.

Redundancy

There's not a nice way to say it. That's boring! This example illustrates the first thing not to do. Writing about the analysis does not mean that you repeat line for line what is in the tables. The reader can easily see in the table what the author has written in this example.

Typically one does not write about the sample size (*N*) in the Results section. Only rarely is it important for understanding the results. If sample sizes are small (less than 100 as a rule of thumb), it may be important to mention to the reader that the statistics being used are more vulnerable to the influence of individual cases (outliers) because the sample is small. In this example, however, this is not the case, and the comment about the sample sizes is redundant and wasteful.

Meanwhile, the author obviously wants to draw attention to differences between the groups. So it may be better to simply point out that there is a difference of X dollars between the two groups' average incomes without pointing out the raw values from which this difference was computed. Readers can verify your math if they want to, but at the same time the author can efficiently point out a difference that he or she thinks is important. Another way to do this would be to point to the percentage difference between the two groups, identifying the fact that the older group makes about 30 percent more than the younger group.

The student's first draft, shown above, is equivalent to the museum guide's pointing to the *Mona Lisa* and saying, "Notice that she has long dark hair, has fair skin, and is smiling." The guide's statement is so boring, obvious, and redundant that it might even be insulting to the museum visitor. Consider how much better it would be for the guide to say, "It is not her dark hair and fair skin that draw attention but rather her mysterious smile." Here the guide points out some potentially overlooked characteristics and helps the listeners move quickly to consider the most important.

Let's consider a revision of that earlier focus on Table 9.1 to try to achieve the same confident mix of observation, description, and a bit of interpretation.

> Table 9.1 demonstrates that mean income is dramatically lower for younger men than for older men. Older men have incomes about 30 percent higher than young men's. This is consistent with Oppenheimer's (1982) argument regarding the "life-cycle squeeze" wherein younger men obtain lower annual earnings at the start of their careers. Meanwhile, the mean is higher than the median for both variables among both groups of men, suggesting that significant outliers are inflating mean earnings and incomes. Even in the upper end of the income brackets the differences are pronounced. The upper quartile of older men make more than $55,000, but for younger men, the upper quartile begins at $43,000.

Notice that with this revision, the author still points to the important issues but helps readers begin to interpret the numbers by (1) making comparisons between younger and older men, using simple words such as *lower, higher,* or *but;* (2) using descriptive words such as *dramatically, inflating,* and *pronounced;* and (3) avoiding a dull rewriting of all the numbers but highlighting only a few specific numbers. The author begins to offer some explanation for why they appear as they do, without getting bogged down in the theoretical explanations for these findings, but also makes an insightful observation about how these findings are consistent with what other authors have anticipated. This revision provides readers the freedom to read the table for themselves and to consider what the author wants them to begin to conclude.

Voice: Visible or Invisible Authors, Analyses, and Audiences

This revised paragraph raises the issue of voice. That is, how visible should the author and the author's analysis be in the presentation of the results? The first draft shown above begins with "My analysis . . . ," bringing the author onto the stage by saying "my." The author also highlights that the analysis is showing something, as opposed to the data's revealing something. These are issues of taste and editorial license, but they are important because at times the author and the analysis can get in the way of the results and findings.

Notice in the longer example of analysis of Table 9.1 that the revision has moved the author and the analysis off of the stage by sidestepping ownership of the analysis ("my"). In this revised example, the table of analyzed data is the source of authority and information rather than "my analysis."

Let's consider a further example. Using Table 9.2, the author wants to show that the educational level of men has an important effect on their incomes.

Table 9.2 Income by Age and Education for American Men

	Mean	*Median*	*n*
Young men (18–40)			
Less than college	30,040	26,000	7,786
College graduate	53,483	42,000	2,981
Older men (41–65)			
Less than college	37,507	32,000	7,246
College graduate	69,129	51,421	3,641

Here is a first draft of some text about Table 9.2:

> From Table 9.2 you can conclude that educational level is extremely important for increasing the income of all men, regardless of age. Computer analysis of the data also reveals that the effects of college graduation grow over a person's

> life. While among young men there is a $23,000 difference between college graduates and less educated men, this gap grows to $32,000 for older men.

This text raises two more issues of voice and the visibility of author and reader. The author says, "you can conclude." The familiar *you* finds no place in formal academic writing. This is because the meaning of *you* is ambiguous. Is the author implying that the reader needed his or her permission to make this conclusion? Is this a veiled invitation to make this conclusion? A command to do so? Is this conclusion optional, such that some could make such a conclusion and others could not? Hence the ambiguity. Here is a related case:

> In looking at the average income of men within different educational categories, we can say education has an important impact on.

When the writer says, "we can say," there is an assumption that the reader will want to say it too. The reader is once again visible but now is being asked to join the writer in saying something. The following approach would work better:

> The observation that income varies so widely with education supports other researchers' claims that education is one of the most important influences on incomes.

This revised text puts the responsibility on the author to assert the meaning of the data. The text does not ask the reader to say it too but allows the reader to accept or reject the interpretation offered.

Consider again the earlier version of the text focused on Table 9.2, focusing on the second sentence:

> Computer analysis of the data also reveals that the effects of college graduation grow over a person's life.

This text also brings the computer onto the stage. Generally, this is unwise. The reader does not really care whether the author computed these statistics on a computer, an adding machine, an abacus, or the back of an envelope. Similarly, references to computer software are generally not required (e.g., "analysis of the data with SPSS"), although you may occasionally see published research where the authors believed that the software's unique abilities needed to be highlighted. However, in general, it is best to let the computer be invisible.

Because most computer programs cannot handle names of variables like "Men's Earnings," they use truncated names like "MENSINC." Do not use these computer-generated code names in tables and/or in writing about tables. Readers should not have to learn a new vocabulary to read the Results section. Even if the computer prints out attractive tables with MENSINC as the heading of a row or column, change this back to its real meaning, and discuss it as such in the text.

Consider this side point: For researchers and students who have struggled with completion of their analysis, it is tempting to communicate to readers how hard they worked to produce this analysis. For example, an author might want to say, "Painstaking and time-consuming efforts to compute the differences in earnings demonstrate that indeed . . ." Unfortunately, the readers of academic writing are not interested in the difficulties of research. Indeed, the author's task is to make the results seem so self-evident and self-revealing that the reader will believe that these results effortlessly presented themselves. This observation stands in contrast to the kinds of information that a tour guide would provide: We actually find it interesting that the painter completed the portrait under difficult conditions.

So who should be visible and invisible in writing about results? For sure, the computer and the readers should be invisible. The data or the analysis can be visible, although the author should beware of putting excessive focus on the analytic process and keep attention on the results. And the author? This remains a point of disagreement among academic writers. In the revision for Table 9.1 suggested, the author remains offstage and simply makes statements about the results, letting them be the source of authority and information.

Some writers stand on the stage with their analysis, introducing each phase of the analysis almost like magicians who say, "Next, I pull a rabbit out of a hat." For example, the author above could introduce Table 9.1 by saying, "I first compute the mean and median earnings for both groups of men. Table 9.1 demonstrates that . . ." Thus, the author takes a more central role in the presentation of results and writes in the present tense. However, notice that the table is still the source of authority and information. In large part, the choice of whether or not the author appears in the text, usually as "I," is an editorial choice that will meet with approval by some readers and disapproval by others.

Tense Yet?

In all of the weak and strong examples provided so far, the author writes in the present tense. For example, "Table 9.1 demonstrates . . . ," and

"Computer analysis reveals . . ." This may feel somewhat awkward to the author since the results actually have been created over time through a laborious process of data construction and analysis. As a result, many first-time researchers are inclined to write in the past tense something like this:

> Evaluation of the data revealed that the gap in earnings between the two groups of men was very large.

Authors of articles published in many social science journals write in the present tense when discussing quantitative analyses. This is true even when they are writing about aggregated data covering several decades! The rationale is that if the analysis revealed something last week or last year, it reveals the same thing today. So it's not that Table 9.1 said something just on the day that the statistical analysis was completed; the results continue to say the same thing. The reader can recall that the data were collected during a certain time (this information is revealed in the Data and Methods section), and the date on the paper indicates when the author is making the current claim.

The benefit of writing in the present tense is that it makes the quantitative results more compelling. However, some social science journals and some professors prefer that you write the paper in the past tense.

It should be noted that social science research based on participant observation or face-to-face interviews may best be communicated by writing in the past tense. If the research process is integral for understanding the results, then this particularly makes sense. For example, if the researcher wants readers to know that the setting in which the data were collected may have influenced the findings, then it makes sense to say so.

> Almost 75 percent of the workers indicated that they were not being paid enough for their work, although when the boss entered the room they quickly changed the subject and hid their questionnaires.

or

> I pressed the managers for more detail when they evaded my questions about the earnings of workers down on the shop floor.

In these instances, the data and the acquisition of the data require that the author write in the past tense. However, quantitative data are generally treated (perhaps naively so) as timeless and context independent, and thus academic writers most often talk about them in the present tense.

Directing Attention to Tables and Graphs

Consider again being on a tour of a museum during which the guide repeatedly drones, "Look at this painting—it is called . . ." At some point you would begin to wish that the guide would quit saying, "Look here, look there," and instead simply point and start talking about the different paintings:

> Compared to the *Mona Lisa* in the other room, the portrait of her sister here looks quite different.

In the same way, it is challenging to point out tables and figures without being heavy-handed. Here are a few examples from some students' writing about tables and figures:

> Looking at Table X for men's earnings and focusing on the mean and median and comparing . . . , it shows that the mean and the median are . . .

> Table X illustrates that . . .

> The data show that . . . (Table X).

> Consider Table X, which shows that . . .

The first example contains an implicit command to join the author in looking, and the next two just assume that readers will want to verify what the author is asserting, with the author merely pointing out the location where this verification can be found. The last one commands readers to look for themselves. One or two of these commands may not be bothersome to readers, but many of them will make readers feel like they are being bossed around. The goal is to focus on the findings by either stating what a certain table or figure reveals or by using the parenthetical maps (e.g., Table X, Figure X) to point people in the right direction for confirmation of the claim.

Earth-Shaking, Surprising, Considerable, and Negligible Results

> "The chances of becoming homeless increase *an astonishing* 25 percent for people who suffer from . . ."

> "By controlling for . . . , the gender gap in pay *plummeted* to only . . ."

> "The impact of . . . on . . . is *highly significant.*"

"The average total income (using the mean) was *much higher* than the median."

"There is a *real discrepancy* between the average income of highly and less educated men."

The Results section of the paper is the first place where the author can begin to provide some interpretation about how surprising or expected the results are. After many weeks of painstaking work, the temptation is to claim that the results are remarkable or awe inspiring when in fact they are much more modest. On the other hand, many authors are excessively humble and fail to assert the importance of their findings. This is where colleagues and reviewers are helpful for determining how big or little, important or trivial, memorable or forgettable, significant or not are the results of the research.

Phrases such as *much higher* and *real* are open for argument. *Significant* may mean statistically significant or substantively significant (whether or not we tested its statistical significance). Beware the apparently neutral phrase *much higher.* There is definitely a place for being persuasive and honest about findings, and if the difference is huge, noteworthy, or much bigger, then say so. But make sure that you keep in mind the cynical reader whose first reaction might always be, "Oh yeah? Why do you say that?" If you can defend against such an aggressive reading of your work, then you are right to give emphasis to your findings.

In one of the preceding examples, the author has indicated that the difference between the two groups is "real" (an apparently reasonable and testable assertion of statistical significance). Words such as *big, real,* and *important* have their place in a Results section, but be prepared to defend them. Consider how they either might be misunderstood or might raise red flags for the reader.

Final Thoughts

Writing about quantitative data is one of the least common experiences for most social science majors. It is hard to do well, especially the first couple times you do it. You have numbers and tables that might be boring to some people but that tell an important sociological story. Overcoming the dullness of numbers and tables to appropriately reveal the compelling story behind them is the challenge. The final figures and tables represent hours of hard work, so it is difficult to remain understated and casual enough to keep yourself, your computer, and your painful research experiences off of center stage so that the data can tell the story. Yet the data do not really tell the story on their own. You are the tour guide who must help the reader to see the whole story in the data.

Chapter Ten

Quantitative Papers

Discussions and Conclusions

Sometimes the Discussion section is separate from the Conclusion section, and sometimes these are the same. You will need to consider where you expect to present your paper and find out the structural requirements of the editor, professor, or agency that will be reading it.

For now, we'll assume that these elements are in the same section. Let us consider (1) the content of such a section and (2) the tone of such a section. To do so, here is a segment of a Discussion/Conclusion section from an early draft of a real manuscript submitted to a journal (Edwards 2002). This paper concerns the effects of education on women—specifically when they give birth to their first child. The paper looks only at those who do become mothers. The main idea is that education can delay motherhood for several reasons: One is because most female college students choose to remain childless while completing school; another is because having invested in education, they want to cash in on that investment; and third, they may be in higher-prestige occupations that they find difficult to leave or interrupt because of motherhood. The findings are reported earlier in the paper. The author must now recount what has been found and begin to make sense of the findings. Read this, and then consider the comments below about what the author is doing:

> The importance of role-incompatibility effects, the absence of educational investment effects, and the modest evidence of occupational effects point out the value of focusing attention on mothers and isolating the theoretically distinct components of education's effect on age at first birth. While many things have changed that make later age at first birth a reasonable strategy

> for American mothers to satisfy parental and occupational ambitions, there is much that remains the same. One year of education generally accounts for one more year of fertility delay.
>
> Education does not appear to be regarded as an investment which will bring a return, neither in 1969 or in 1987. Other than women in professional, technical, and managerial occupations, employed mothers delay first births about the same length of time within any given year. Meanwhile, very large shifts in the occupational distribution or in educational attainment do not help explain the pervasive trend in post-education fertility delay. Indeed, women in higher prestige occupations delay their first births longer, but they do not comprise a large enough component of the entire occupational structure to significantly affect the overall trend.
>
> Why might we observe no evidence of educational investment effects while also locating greater fertility delay for women in higher prestige occupations? In spite of a three-fold increase . . .

Notice that in the first line the author is trying to sum up the findings (e.g., "importance of," "absence of," and "modest evidence of"). And he is trying to remind the reader that this evidence makes clear that his method ("focusing attention on mothers") and theoretical contribution ("distinct components of education's effect") are something that he has done and others have not. The author promised this earlier in the paper and now is reminding the reader that he has delivered.

Next, the author acknowledges that there has been a big trend and that he has not explained it all. But he also sets up the reader to see that the findings locate more sameness than difference over time. He is trying to dramatize that which he has found and that which readers would not have expected. Then, he restates the specific findings that illustrate his point. In the same way that we must be cautious in overstating the results (discussed in chapter 9), we must be accurate, confident, and cautious in the conclusions we draw. The author is trying to do that with the things he asserts he has done and the admissions about what he has not done.

The next paragraph draws attention to what was a subpoint in the analysis yet what the author thinks needs to be reasserted. Notice that he acknowledges what we would have expected ("women in higher prestige occupations delay their first births longer"), but he wants to emphasize his point that this group is not large enough to account for the overall trend.

Finally, he poses a rhetorical question to set him up to theorize (some might say "speculate") about why his findings turn out how they do. He goes on here to suggest some theoretical possibilities rather than complain about the inadequacy of his measures.

A Few Words about Tone

When you write your paper, as when you are talking to friends and family, how you say something matters as much as what you say. (Recall teachers or parents who told you, "Don't use that tone of voice with me!") Your paper will have a tone to it just like your voice does, and it will have a tone whether you attend to it or not.

The arena in which you are writing should influence the tone, or the style of voice, with which you write. A U.S. Census Bureau report rarely takes on an aggressive, assertive, or even partisan tone. Such a report tends to be very detached and reports just the facts. Meanwhile, contrary to what we might think of academic journals' being boring and noncommittal, most papers actually assert and press issues near the end of the paper, even though they do so in a diplomatic and restrained way. If you were writing a report for your social service agency and needed to demonstrate that in fact the population of people you are serving really needs more attention or funding, then you might press your claims more dramatically and forcefully—although your decision to do so would depend on your strategy. Sometimes, the facts speak for themselves. Sometimes, you need to highlight, contextualize, and dramatize them.

Some Examples of Tone

In the writing sample considered earlier, the author seeks to establish a tone of quiet confidence, respectful disagreement, and humble acknowledgment of what this paper does and does not accomplish. He nods in the direction of some critics, ignores others, and tries to keep his attention on the analysis without worrying too much that he did not think of every possible exception to the rule.

Look again at that text. The author uses words such as *importance, value,* and *distinct* in the first line to emphasize confidence in his findings. He uses words such as *modest* to indicate that there is something there but that it is smaller than we might have thought. When he says, "while many things have changed," he is agreeing with potential critics that there is a lot more to the story. But then he also goes on to make his point about sameness over time.

In the last part of the first paragraph the author somewhat curtly restates his points (just like pointing to the irrefutable evidence such as a smoking gun in a court case).

Finally, the author uses the rhetorical question to establish a sense of ponderous reflection, calling readers to join in careful consideration. (There's a risk here that readers will feel manipulated by his saying "we," but apparently he was willing to take that risk.) The writer is hoping that readers will respond by saying, "Yeah, I was wondering about that . . ." In this case the author is trying to set a tone that he thinks fits with the arena (an academic journal) and that will resonate with his readers (academics and students).

A Last Example

Let's consider another example of a conclusion from a paper—this time from a final draft of an undergraduate student paper. (The student gave permission to include the paper.) Although it is the final product, it too could use some significant editing to make it stronger.

> The findings for this study strongly support the initial hypothesis that among single mothers of preschoolers, the mother's educational level plays a substantial role in shaping her participation in the labor force. It is observed that mothers with at least a minimal amount of education beyond high school are much more likely than those without post-secondary education, to be employed in the labor force (Table 1). For example, among those lacking college experience, nearly 67% are unemployed, while among the mothers with from one to five years of college education, almost the exact reverse is the case with 66.4% of them represented in the paid labor force. In addition, the Pearson's chi-square test result (shown in Table 1) verifies the statistical significance of the observed relationship at the .001 level. . . .
>
> When the variable (number of children) was controlled for (Table 4), the positive relationship between educational attainment and labor force participation was not significantly affected. The same pattern that was seen initially in the original observation emerged again in each sub-section (i.e., both "mother with only child" and "mother with 2+ children" replicated the initially observed relationship). Among mothers with only children, college education seemed to have a bit less effect than was shown in the initial relationship, in that there was a greater percentage of mothers with and without college educations in the labor force than the initial table had reflected. . . .

Here are some observations about this text.

a. The text excessively restates the specific findings. Indeed, we may wonder how this section differs from the Results section. It would be better to focus instead on the presence of an initial relationship, as expected, and the fact that it remains fairly stable even when the control variable is controlled.
b. References to tables and particular statistical tests are only rarely appropriate in the Discussion/Conclusion section. These highlights should have been made in the Results section; here we should talk about them a bit more like they are general findings ("more highly educated women are more likely than less educated women to . . .").

Together, items a and b indicate that this student is a bit caught up in the method and what has been accomplished and is not free enough of the analysis to tell the story that the data tell. What about tone?

The tone here is certainly not pushy or flamboyant. The author uses passive voice to communicate (e.g., "It is observed . . ."). He asserts that some relationships are "strongly" supported and that some variables play "a substantial role." The excessive wordiness of the sentences and attention to so much detail end up making the tone of this heavier and more confusing than it needs to be. Here is a suggested revision:

> As expected, for single mothers of preschoolers, having any education beyond high school increases the chances that they will be in the labor force. Indeed, while only one third of the single mothers with no college experience are employed, about two thirds of those with college experience are employed. This relationship persists among women with only one child and among those with more than one child. However, for mothers of only children, college seems to have slightly less effect than for all mothers combined.

This is not the only way to revise the student's text, but notice that the essential elements still appear and leave open the space now to discuss what it is that might help us understand why education plays such an important role for single mothers (as opposed to married mothers).

A Closing Argument

Let's return to the courtroom scene with which we began. In any good courtroom drama, the closing argument is when the details are summed up for the jury, the strengths and weaknesses of the evidence are described, and a verdict is requested. The Discussion and Conclusion sections of our research papers are not nearly so exciting as the emotional appeals that seem so persuasive in television courtrooms. But they are the place where we finally say, "This is what you should get out of this paper."

Solid Discussion/Conclusion sections of research papers find a middle ground between extolling the paper as the slam-dunk answer to all important questions and apologizing for the paper. Novice authors tend to fall into either of these extremes, perhaps hoping that bluster will keep the reader from remembering the weak parts (e.g., "This paper has finally resolved the debate between . . ."), or hoping that by confessing all the paper's weaknesses, the author can distance him- or herself from the paper enough to avoid criticism (e.g., "This paper is only a small-scale pilot study investigating ill-defined ideas . . ."). Whatever the psychology, recognize that the final words of your paper are as important as the first and subsequent words. Go ahead—draw appropriate conclusions, and state why they are trustworthy and important; acknowledge what is not yet clear, but don't tear the paper apart so badly that readers wonder why you ever asked that it be read. Your paper, well constructed, has something to say!

Reference

Edwards, Mark Evan. 2002. "Education and Occupations: Re-examining the Conventional Wisdom about Later First Births among American Mothers." *Sociological Forum* 17(3):423–43.

Section III

Writing Qualitative Papers

Chapter Eleven

Ethnographic Interviewing and Storytelling

The following paper by Rich Mitchell and Kathy Charmaz does a great job of showing you how self-aware writers think about using qualitative data to tell powerful sociological stories.

Mitchell, Richard G. and Kathy Charmaz. 1996. "Telling Tales, Writing Stories: Postmodernist Visions and Realist Images in Ethnographic Writing." *Journal of Contemporary Ethnography* 25:144–66.

TELLING TALES, WRITING STORIES

Postmodernist Visions and Realist Images in Ethnographic Writing

RICHARD G. MITCHELL, Jr.
KATHY CHARMAZ

PROLOGUE

This article is a conversation between two authors concerned with writing ethnographic tales in interesting ways. Ambivalence, multivocality, and give-and-take of perspectives are intended parts of this discussion. To begin, Richard Mitchell writes of his experiences in the field and as an ethnographic storyteller. Then Kathy Charmaz writes about the analysis of stories, beginning with Richard's account and generalizing to other ethnographic tales. Richard responds with an addendum to Kathy's analysis, stressing writing as an experience in its own right. We conclude where we began, in the field, with a second look at phenomena, storytelling, and authorship.

Ethnographers and writers in other genre rely on similar techniques (e.g., Bickham 1994; Golightly 1970; Noble 1994; Oates 1970; James 1989; Krieger 1984; Provost 1980; Wright 1989; Yolen 1989). They employ five basic strategies: (1) pulling the reader into the story, (2) recreating experiential mood within the writing, (3) adding elements of surprise, (4) reconstructing the experience through written images, and (5) creating

---

AUTHORS' NOTE: Richard G. Mitchell thanks "John Huntley" for speaking his mind and for his timely gift. Kathy Charmaz thanks Gerald Rosen, David Bromige, and William De Ritter from whom she may have learned more than any of us realized. We both appreciate the comments of Norman K. Denzin, Mitchell Duneier, Lyn H. Lofland, Laurel Richardson, and the editors of this issue. Portions of this article were presented at the 1995 meetings of the American Sociological Association, session on Qualitative Methodology.

JOURNAL OF CONTEMPORARY ETHNOGRAPHY, Vol. 25 No. 1, April 1996 144-166 closure on the story while recognizing it as part of an ongoing process.[1] Good writing reflects these strategies: they unify a work and move it toward its conclusion. That is part of our message. The other part is a caveat. Good writing transcends technique. These strategies are sensitizing, not prescriptive. Concern for technique alone does not help us understand writing, good or bad. Writing like all forms of knowledge is ultimately intuitive, not methodical (see Sartre [1953] 1966, 240). Consider the following example, "The First Interview," an excerpt from the beginning of Richard's twelve years of study among survivalists. Survivalists are people who take seriously the possibility of imminent social disruption—economic collapse, foreign invasion, even antigovernment violence, and internal race war.

THE FIRST INTERVIEW

Recalling that first interview brings immediately to mind the March 1983 issue of the White supremacist publication *The National Vanguard.* It was a gift from my first informant. It came at exactly the right moment.

I had just finished reading "Survival Treasure Chest: Today's Pieces of Eight Make Sterling Investment" (Kogelschatz 1983, 38-39, 41) in *Survive* magazine, to which I had recently subscribed. Gold and silver were the things to have in times of crisis, the article advised. No survivalist should be without a good supply. If this were the case, I reasoned, then my hometown precious metals dealer John Huntley[2] might be in contact with local-area survivalists and, therefore, his shop would perhaps be a good place to start my research. I called him for an appointment. As it turned out, I was right and I was wrong. Huntley knew about survivalists, but not the ones who hoard gold and silver.

I armed myself with a notebook and a list of naive questions ("How many survivalists would you say visit your shop in a typical week?" "Approximately what proportion are males, females?" "What would you estimate the average age of these survivalists to be?"). Feeling a bit insecure about my reception,

I concealed my already-running tape recorder in a light fabric bag. I had forty-five minutes before the recorder would reveal itself by a loud end-of-tape click. I walked into Huntley's shop, shook his hand, and accepted the chair he offered at one side of his desk. Arranging my papers, and setting the concealed tape recorder on the desktop, I was ready to begin.

I introduced myself. I explained that I was a sociologist working on a book about survivalism and would like his help. I showed him the *Survive* article on investments. He had never seen the magazine before. After looking it over, he opined, without much interest, that to his knowledge no survivalists of the sort the article described came to his shop. Fifteen minutes passed in unfocused talk. It seemed Huntley thought little of stockpiling gold and silver as a survival strategy, or of my writing project.

Then the conversation took an unexpected turn. Our roles reversed. Huntley began to ask me questions about myself, my wife, and the nationality of our parents and grandparents. When he learned my wife and I were both university faculty, and of German and Norwegian descent, he grew excited. "You talk about survival," he said, "I've made an in-depth study of that. . . . You and your wife would be prime candidates to be taught the realities of the last hundred years of United States civilization, and what's going to happen to us if we don't wake up!" Huntley had his own apocalyptic vision. And he had something else I just then noticed. On the desktop to his right, partially covered by a few sheets of paper, lay a .38 caliber revolver pointed my way.

The future looks grim, Huntley asserted. "We are living in a collapsing civilization. It's like an implosion." The cause of this failure? "The cultural bearing stock, the Anglo-Saxons and northern Europeans that are the problem-solving peoples of our civilization are being displaced." The presumed best of this lot, the Nordics who "fought the Mongol hordes in Europe . . . (and later) got in covered wagons and came west and survived in the wilderness" are at special risk. According to Huntley's reading of census publications, "Nordics are only having about 1.2 children per family." He added,

> White people instinctively know things are wrong, particularly the Nordic because he is very sensitive to his surroundings. Even though he may not be able to verbalize this distinctive feeling, he stops reproducing, especially in the big cities. It was the Nordics that built New York City, but it's a fact that the ones who live there now have literally stopped having children.

His prognosis: "If you project over a 150-year period in the United States, the Nordic will be extinct!" And elsewhere, "It's worse, it's worse in Sweden, Luxembourg, in France, even Russia and the White Communist countries, Poland, Czechoslovakia, Hungary." Other hazards are more immediate than extinction, Huntley warned.

> We were 90 percent of the population up until the Civil War, now we are probably only 60 percent. When we become less than 50 percent of the population, then, living in a democracy, the other groups are going to completely dispossess us. We will be the dispossessed minority.

Huntley added details, examples, citations from Charles Darwin and George Orwell, from Marx and Hitler. The tone of the conversation changed again. He began to speak of "us" versus "them," to include my wife and me in his cause. He moved his chair closer, leaned forward. His tone became conspiratorial, as if secrets were being shared. "I'm not interested in giving you this information just so you can write a book," he clarified, "but for your own information. Then you can do with it what you will, because you might become a recruit. Then you will go out and want to proselytize." He seemed anxious to incorporate me into this hypothetical fate, to stress the personal seriousness and urgency of the Aryan's problem. At that moment I faced another problem, also serious and urgent. I glanced at my watch. In nine minutes, more or less, Huntley would discover I had been tape-recording this conversation. The trust he seemed to imagine existed between us might suddenly end.

Right then I wished I had known more about the man. A few things were apparent: his clear blue-gray eyes, thick dark hair, athletic build, and clean, delicate hands. Some facts about him, had I known them at the time, might have put me at ease. He was 44, married, and had two teenage children. He held B.A.

and M.A. degrees in music and had done some work toward a Ph.D. He had been a schoolteacher and was once an unsuccessful congressional candidate. However, other information, also obtained later, would probably have made me even less comfortable.

As it turned out, Huntley was well-known to area journalists and local government officials for his frequent, unsolicited essays, phone calls, letters to the editor, and speeches at public meetings.[3] The timing of these expressions of opinion was unpredictable, but the themes were consistent: international Jewish conspiracy, growing government repression, impending "patriotic" rebellion. Huntley was on record as having claimed, among other things, that "the Illuminati hired Marx to write the Communist Manifesto"; the U.S. military is preparing to quell a "nationwide tax rebellion of six million people, maybe more, who aren't even filing any tax returns"; the government has readied Operation Cable Splicer, which will "isolate various areas of the country [by creating] power blackouts and various communications breakdowns, then move the military in, . . . [confiscate guns and property and] arrest those people they consider dangerous—like myself." But "when they try it," Huntley bragged, "we will kick their butts right out of the country. I don't know what's going to happen but there's going to be bloodshed!"

Huntley predicted that at the head of this "patriotic rebellion" will be the Posse Comitatus, in which he had long been active. The Posse Comitatus, I was to learn, advocates armed resistance to what they view as illegitimate taxation based on personal income or property, and most governmental authority superordinate to the county sheriff. In retrospect, this would have been relevant background material and certainly would have helped me understand what was going on, but I knew nothing of it at the time. Instead, I could only continue to listen, unsure of the interview's course, or outcome. The tape recorder kept running.

From a file cabinet next to his desk Huntley brought out the then-latest edition of *The National Vanguard*. He was quiet for a moment, glancing through the issue as if to remind himself of its contents. Then, placing the magazine on the desk between us, he continued.

"They talk about racism," he said. "Well, I'm a racist. I believe in preserving all the races, but not mixing them together." Huntley believed some social science, wittingly or not, contributes to the denigration of racial purity.

He continued,

> One of the great misconceptions that the American civilization has been under since World War I has been the egalitarian or the equalitarian philosophy which is spread through the Franz Boas and Margaret Mead school of anthropology. . . . When he [Boas] came to the United States, he was thoroughly imbued with Marxism, and the whole basis of Marxism is egalitarianism. In other words, you cannot admit to racial or individual differences if you are a Communist.

Huntley argued that this ideology of egalitarianism, although not part of the Constitution, has come to permeate educational curricula and governmental policy. "What you have here is Marxism. It goes from the very highest echelons of our federal establishment right into the school system." Marxism is only a symptom; it is not an end in itself, Huntley explained, but the means by which international Jewry seeks to gain control. "What the Jews want to do is reduce us to the lowest common denominator, not just socially but biologically. They want to destroy the cultural bearing stock." Current social policies further this end, Huntley argued.

> That's what integration does, it mixes the gene pool. It destroys the cultural bearing stock. And look at the manipulation in this zero population growth. The only people that have cut back on their population are the Europeans, especially the Nordics. Your Blacks and Mexicans and Vietnamese and other ethnic groups keep right on breeding.

Both Huntley and I had become agitated. Huntley seemed to care very much about the issues at hand. As he had told an earlier interviewer, "Once you get into this, if it piques your interest, you'll never get out of it. You just dig and dig and dig until it consumes you."[4] But Huntley seemed to be enjoying himself. Here he had, at once, an apparently receptive audience, perhaps a potential recruit, and the chance to unveil what he saw as a fundamental but overlooked principle of social science to a credentialed sociologist. In contrast, I felt confused and disoriented. Huntley was an obviously intelligent, widely read, articulate individual, a resident of my own community, yet he espoused racism of a sort I believed would be found only among bucolic bumpkins or the genuinely demented. The interview had drifted far from my intentions or control, my liberal sentiments had been summarily rejected, and my ability to withhold judgment was growing frail. Fieldwork was proving more than I was prepared for. In my uncertainty, I said little. Whereas Huntley was anxious to reveal what he knew, I was trying to keep a secret. And frankly, I was frightened. If another's ideas could be so contrary to my expectations, what then of his behavior? How would my dishonesty regarding the tape recording be received?

As Huntley spoke, I thought of the time and of his impending discovery, now no more than a minute or two away. Huntley grew even more animated. Leaning forward again, forearms on the desk, his right hand strayed toward the revolver, brushing the papers away. Like an engrossed thinker stroking his chin he began idly to rub his palm back and forth across the gun's cylinder. A sliver of afternoon sunlight breached the shutters and glinted off the gently swaying barrel. From my muzzle-on vantage point, and in this direct light, I could see the chambered bullets were a copper-jacketed, hollow-point design.

It ended as T. S. Eliot predicted the world will, not with a bang but a whimper. As a final punctuation to his discourse on yet another topic, the misrepresentations of "Black History Month," Huntley picked up *The National Vanguard*. "If you want to find out what is really going on, read this!" he enthused, slapping the publication down on top of the concealed recorder. The "whap" of the descending magazine and the "click" of the ending tape coincided. Almost exactly. The whimper was mine, a partially suppressed, involuntary cry of fear and relief. Puzzled by my utterance, Huntley offered reassurance. "You can keep it if you want," he said, tapping the journal. I thanked him, assured him I would look it over carefully and consider what he had said, and gathered my things to leave. The first survivalist interview was over. I had come confident of my will and skill to "win" respect and "acquire" information (these were the terms I used in preparatory notes to myself). My presented self, or so I imagined, was that of the competent, objective, purpose-filled researcher. I left in ambivalent confusion—titillated by flirting with apparent danger, befuddled by my naïveté, and frightened by my potential new identity, by what I had been taken for, and might become, if this study continued—a racist.

KATHY CHARMAZ'S WRITING CRITIQUE

What can we learn about writing ethnographic tales from Richard's story? Through studying his story, I offer some practical guidelines for turning research tales into written stories. These guidelines are for qualitative researchers to consider, to play with, and perhaps to adopt or revise. Analyzing a story reveals most clearly how the guidelines work. However, they can be used to enliven less dramatic ethnographic description and more explicitly analytic works (Charmaz 1991). Although my analysis emphasizes finished stories, I recommend adopting these guidelines as observational and writing strategies from the beginning of the research process. The guidelines are tools. Nothing more, or less. They may help us observe more closely, write more gracefully, and thus state our ideas more artfully.

PULLING THE READER IN

What induces us to read a story, article, or book? The opening paragraphs or the opening chapter should pull us into the story and convince us to continue. Hence the writer invites, entices, and involves the reader to stay with the story and to remain in the scene (see Hubbard 1988; Noble 1994). To bring us into the story, the writer needs to provide its context or to imply what might follow. Often qualitative researchers use a telling anecdote, case material, or an interview excerpt to do just that. A telling opener piques our interest and curiosity. In my own writing, I often focus on a concrete person or specific incident to stimulate reader involvement in more general themes.

A carefully selected opener allows the writer to make implicit or explicit claims from the beginning. Writers who retell their intense experience, rather than recount someone else's, recreate its power through their written images. In more formal writing, I look for a clear, spare opener in the first paragraph or two, which states concretely and specifically what research or analytic "story" this work will tell. When the author's thesis is general, the problem common, or the argument unclear, I lose interest. You probably do too.

In "The First Interview," Richard accomplishes four writer's objectives in his first short paragraph. He (1) identifies the viewpoint of the story, (2) persuades the reader to become intrigued and emotionally involved, (3) sets the mood, and (4) hints of suspense and conflict. From the start, we know that the story proceeds from his viewpoint. Told from any other viewpoint, it would not be the same story. Richard brings his readers, especially other social scientists, right into the scene with images of the White supremacist magazine. By mentioning that it was a gift from his first informant, he taps others' images and memories of initiating fieldwork. Thus he establishes a common ground with us, his readers. But he pulls readers into the story and keeps them reading and wanting to know more when he states (about the magazine), "It came at exactly the right moment." Why was it the right moment? In this brief cue, Richard sets the stage for telling his tale. Similarly, he develops the scene when he talks about starting his research and making contact with Huntley. He arouses our curiosity further when he says, "As it turned out, I was right and I was wrong." What happened next? How did being wrong shape later events?

We want to know what happened. Richard's writing style establishes a personal connection with us.[5] Through his use of language, imagery, rhythm, order, and authentic voice we imagine a whole human being who lived the story rather than hear an anonymous report of it. His informal style and judicious self-disclosure allows our intimacy with him to grow. I say "judicious self-disclosure" intentionally.[6] Otherwise the writer intrudes and the writing grates. Neither gratuitous inclusion nor intentional omission of the writer's presence (as the positivists would have it) leads to good writing. I find a built-in tension here: the writer is at once the source of meaning and the source of its obfuscation. The extent to which the writer's presence should be central and explicit depends on the nature of the research tale. Ultimately, the effectiveness of the writing partly turns on how the writer handles this tension.

Richard's story presents an interesting case of the writer's subjectivity because he is central to the drama. Yet his voice in telling the tale allows us to understand the emerging events. As Richard tells us what he thinks, feels, and does, he brings us into a jointly felt scene and prompts us to empathize with him (see Nash 1989). Even though the story is Richard's tale, he does not dominate it unnecessarily. He brings himself into the story when needed to move the story along.

Richard's place in the story becomes layered and complex. More than being the narrator who provides the viewpoint, he acts within the scene. He becomes the narrator with a secret, the actor who takes his readers into the plot.[7] Two sharp edges frame this plot: his precipitous plunge into an unexpected scrape and his expected downfall upon discovery of the tape recorder. Richard's immersion in the story fits the tale told.

RECREATING EXPERIENTIAL MOOD

Recreating the mood of an experience through the writing keeps the reader engaged. In addition, it unifies the scene and tightens the story. Cheney (1983) describes a pure scene as all action with minimal distractions. Only those narrative details are included that enliven the scene.

As a writer, I think about what kind of mood an experience, event, or encounter reflects and then I write it into the description. If I am working with a more abstract idea, I ponder about how I want to cast it within the analysis. When writing a story, a unified portrayal of characters similarly furthers recreating the mood of the experience and lessens distractions. To do so, a writer may sacrifice efficient writing, that is, narrative description, for an effective story. Thus Richard uses direct quotes from Huntley's diatribe, offers his internal monologue, and provides reflections about the scene while in it.

Richard takes us through the shifts in mood as the story progresses. Our involvement intensifies and our suspense increases. His imagery and candor place us in the scene with him. "I armed myself with a notebook and a list of naive questions." We can all imagine doing this. We identify with Richard as he sets out to play the role of social scientist to a respondent in an unfamiliar setting from whom he expects only preliminary information. As events proceed, we sense growing ambiguity and his waning morale. "Feeling a bit insecure about my reception, I concealed my already-running tape recorder in a light fabric bag." The mood deepens. We feel Richard losing ground, "Fifteen minutes passed in unfocused talk. It seemed Huntley thought little of stockpiling gold and silver as a survival strategy, or of my writing project."

The mood shifts ominously as Huntley takes control of the interview and Richard notices the gun. Huntley's excitement quickens. We feel it in the short, stark sentences. Tension builds through Richard's comments, not solely through the Huntley excerpts: "However, other information, also obtained later, would probably have made me even less comfortable." Richard's hint makes Huntley more menacing. The twists and turns in the encounter keep us riveted. "Huntley seemed to be enjoying himself. . . . In contrast, I felt confused and disoriented." Urgency mounts as Richard fears he cannot escape before the tape clicks off. "Fieldwork was proving more than I was prepared for. In my uncertainty, I said little. Whereas Huntley was anxious to reveal what he knew, I was trying to keep a secret. And frankly, I was frightened."

ADDING SURPRISE

Throughout the story, unforeseen events pile swiftly on each other. We enter a scene with Richard in which ordinary rules and values are discarded. Expectations dissolve. Uncertainty increases. Roles reverse. Stereotypes collapse. Potential threat heightens. . . . Fear escalates. Richard's initial insecurity sets the mood for his later predicament. The story begins with an ambiguous scene with an unexpected opportunity to probe an unknown character's views. Yet the story does not dissolve into a routine event or a mundane tale of the field. The formidable topic, Richard's apprehensiveness, and Huntley's astonishing adeptness in controlling the interview all preclude that. The momentum quickens, suspense thickens. Early in the tale, Richard warns us of dangers to come, "Huntley had his own apocalyptic vision. And he had something else I just then noticed. On the desk top to his right, partially covered by a few sheets of paper, lay a .38 caliber revolver pointed my way."

Consistent with principles of writing fiction (Carroll 1990; Frank and Wall 1994; Giovannoni 1972; Provost 1980), Richard creates tension and adds surprise by recounting a predicament. In less gripping ethnographic stories, we add elements of surprise by revealing implicit meanings and rules, showing taken-for-granted assumptions, defining worldviews, and explicating hidden processes. In Richard's story, the tape recorder, an instrument for enhancing interview recall and, by extension, the thoroughness of the report, paradoxically is transformed into an obstacle and a liability, threatening to destroy trust, to create enmity, to diminish access to other respondents, and even to damage professional prestige. Already Richard feels uneasy and stuck in an alarming encounter. Then, beyond that, Huntley's imminent discovery of the hidden tape recorder spells disaster. Richard works the drama of his tale deep into his sentences, "At that moment I faced another problem, also serious and urgent. I glanced at my watch. In nine minutes more or less, Huntley would discover I had been tape-recording this conversation."

The tape recorder turns into a time bomb ticking toward an explosive confrontation.

Like most writers, Richard does not allow the events to be a total surprise. He foreshadows the surprises. Foreshadowing limits the surprise and defines the obstacle to overcome. "The conversation took an unexpected turn." Richard warns us here that the predictable interviewer and respondent roles had changed. Writers also foreshadow and limit surprises by planting questions. "If another's ideas could be so contrary to my expectations, what then of his behavior? How would my dishonesty regarding the tape recording be received?"

RECONSTRUCTING ETHNOGRAPHIC EXPERIENCE

Why should readers accept the writer's viewpoint? What prompts anyone to trust an ethnographer's rendering of an experience? I contend the writer's presented images must resemble the experience. Although only evocative of the shared experience, ethnographers must strive to represent their subjects' understandings as well as their own (see Mitchell 1993, 41, 54-55; Prus 1995). Richard's tale may portray an extraordinary experience, a world alien to his readers. Other works address experiences that readers may share (Charmaz 1991; Denzin 1986a, 1986b). Readers will compare their experiences to the ethnographer's portrayed images. What helps writers to create works that seem real and true? How can we reconstruct and represent lived experience through our written images of it?

We must *show* our readers what we want them to know. We cannot simply tell them. Nor can we persuade through mere assertion. Richard shows us Huntley's character as well as his own. He produces Huntley as he forms the tale. Huntley's identity emerges through using his statements such as, "'[The government will] arrest those people they consider dangerous—like myself.' But 'when they try it,' Huntley bragged, 'we will kick their butts right out of the country.'" Quoting Huntley reveals his viewpoint, builds another narrative voice into the story, and simultaneously dramatizes Richard's predicament.

Richard's statements about Huntley further reveal Huntley's character. He reports that Huntley "bragged." Huntley did not state, suggest, or hope—he bragged. Richard underscores Huntley's agenda when he writes, "He moved his chair closer, leaned forward. His tone became conspiratorial, as if secrets were being shared." Richard also produces Huntley's emerging identity as he simultaneously chronicles the unfolding events. "Like an engrossed thinker stroking his chin, he began idly to rub his palm back and forth across the gun's cylinder." The blend of ethnographic commentary and direct statements all contribute to the veracity of the scene. Everything in the story serves a purpose.

Throughout the tale, Richard's tone is consistent. His words and images fit the story. We can imagine the scene. He provides sufficient description for us to surmise his predicament and to feel his trepidation. Richard builds on his vulnerability, the uncertainty, the growing ominousness—his urgency. Like a creative writer, he puts feelings together rather than taking them apart. As Hale (1972) observes, the writer's feeling is a method of perceiving. It renders the writer open to his or her subjectivity. Feeling is concerned with secrets, hiding places, and imagined scenes. Richard's feeling forecasts his precarious position and foreshadows the emerging drama. Even his description fits the twists and turns in mood and thus makes the story powerful. No discrepancies in tone exist. He has calibrated his tone and shaped images to mirror the unfolding events.

Effective word choice contributes to Richard's presented image resembling his experience. His writing gives the impression of natural speech (Packer and Timpane 1986; Provost 1980). It is only an impression, an image. Richard talks with us—so it seems. He reproduces the informality and intimacy of natural speech. He does so through describing actors and events in a conversational style. He also reveals his thoughts and feelings: "The whimper was mine, a partially suppressed, involuntary cry of fear and relief." Richard describes the moment as though recounting it to a close friend. He does not replicate natural speech with all its hesitancy, irrelevancy, redundancy, and inadequacy, yet his words read as if spoken. Giving the impression of natural speech echoes the experience and imbues it with verisimilitude.

Richard's interview excerpts are pointed; they distill the experience. We certainly do not receive forty-five minutes of dialogue in this story. Nor do we view all the images that Richard saw during his foreboding encounter. He creates a painting for us rather than a photograph (Charmaz 1995). Writers need to give us the shape, color, tone, order, and form of their stories;

they do not need to provide the entire experience. Instead, they stress some events, minimize others, and ignore still others. Extraneous detail clutters the story and obscures the point. Writers supply sufficient content in distilled form to make their intentions and interpretations understandable and persuasive. Then readers can imagine the action and, likely, empathize with the writer or main character, but they may not be persuaded. To persuade, writers have to offer sufficient evidence to support the credibility of their claims. The best writers balance the least content with the most powerful persuasion.

The rhythm of the words should be consistent with the described experience. Richard's images are lucid and forceful. For example, he creates emphasis and rhythm through the parallel construction of "titillated by flirting with apparent danger, befuddled by my naïveté, and frightened by my potential new identity." Throughout his tale, the flow of sentences echoes the progression of events. The words sound right. Their length, sound, and cadence create movement and forewarn of impending action.

> As Huntley spoke, I thought of the time and of his impending discovery, now no more than a minute or two away. Huntley grew even more animated. Leaning forward again, forearms on the desk, his right hand strayed toward the revolver, brushing the papers away.

CREATING CLOSURE

Richard's ending is at once compelling and haunting. He does not offer the standard closure. No resolution of the conflict. No heroic overcoming of the obstacle. No ingenious solution to the predicament. Rather, a fortuitous coincidence allowed him to leave unscathed. The coincidence stops the immediate suspense but does not end the story. Instead, Richard's reflection closes this story forcefully—the entire piece coheres. His final revelation opens the possibility of another more powerful drama. This tale is but one chapter in an evolving saga (see Ellis 1995).

Like a novel, Richard's closure is implicit from the beginning. He hints. He reconstructs the foreboding mood. He juxtaposes Huntley's expanding persona against his own shrinking identity. His style, imagery, and voice all move us toward the conclusion.

The meaning of the tale comes through in the last twist, the final surprise: we ourselves are vulnerable to the worlds we enter.[8]

RICHARD MITCHELL'S ADDENDUM

There you have it. A story, and a story about storytelling, both neat, pat, finished, and quite misleading if taken at face value. Let us consider the development of "The First Interview." It was certainly not begun as a project intended to illustrate principles of writing. I knew little or nothing of the succinct and useful writing guidelines Kathy pointed out and explained with such care in the preceding paragraphs. Had I known of them in advance, Huntley's story might have taken other forms. Instead, the Huntley tale began, like most other ethnographic writing, as a mundane list of observations and notes, nearly formless, entirely fragmented. For direction I had only a vague personal sense of the word tone and cadence it would be nice but not necessary to achieve. Drama did not assert itself from the field but was sifted, organized, and built up out of a confused mass of quotidian detail and ambiguous feelings over successive drafts and considerable time.

Feelings have been mentioned and writing is an affect-filled experience to be sure. As a would-be author I had feelings in these times. But far from fear, trepidation, and existential angst, this writing period was one of growing personal satisfaction, reassurance, and fulfillment. The memory of clumsy, faltering fieldwork was set aside and in its place came the adventure of tale-telling. Quiet, safe at my desk or in my favorite seat at the local coffee house, I luxuriated in creating with words and recreating in fantasy my own dual characters, the clever writer and the fieldworker to whom exciting things might happen. I reveled in temporary enchantments. Life was mine to transform, to idealize, to simplify. My pen drew imaginary sides, set rules, made action consequent and lasting. At the heart of writing experience are moments of the intense, imaginative actualization I have elsewhere called "flow" (Mitchell 1983, 153-92). Fleetingly, action and awareness merge, the spontaneous "I" joins the socialized "me" in concerted and complimentary effort. Like mountain climbing, chess competition, delicate surgery, and other forms of cathected action, writing at its best demands the full focus of our creativity and skill. In return, we get what we give. Unequivocal commitment yields a full measure of intrinsic reward.

In these comments, neither Kathy nor I implies support for a distinction between so-called realist and impressionist reportage. There is none. No worthy author's writings derive entirely from empirically ungrounded figments of fantasy. All ethnographic stories are stories of some portion of human-lived experience, experience that is eminently real, immediate, concrete, and meaningful to those who live it. Sartre put it bluntly: human-lived experience "is, it is what it is, and it is as it is" (quoted in Solomon 1988, 180). That is clear enough. All ethnographic stories, too, are stories, more or less imaginative, nuanced, and stylistic interpretations of the worlds we study. Quibbles over the ontological status of the truly true and debates over the primacy of one discourse over another serve no useful purpose. The problem of perception, of obtaining consciousness of the world, is not an issue here. Our concern is finding ways for individual consciousness to join the intersubjective, ways to report experience to others and to ourselves.

All stories, including accounts of scientific knowledge, are relative and provisional. All are but temporary way points in the ongoing construction of meaning. William Pierce, the editor of *The National Vanguard*, publishes facts of human character and history upon which Huntley performs his own analysis. Huntley's account of history and current events is the grist for Mitchell's storytelling. Mitchell's tale provides Charmaz with material upon which yet other sorts of analyses are performed. And so it goes. Facts call out interpretations; interpretations become facts. Realities and impressions answer each other, reciprocate. Last one up claims expertise, authorship, but only until the next telling.

However crafty and complete our stories are, they can be no more than tentative offerings, possible ways of telling from among many. In the field, the people we study talk back, resist, bend, reinterpret, and even reject the images, pictures, and conceptions we and others create of them. We can, of course, insulate our accounts from the risk of empirical disconfirmation. We can shift our studies away from the holistic complexity of human social life to the analysis of symptoms and parts—rate data, written texts, audio- or videotapes, and other ephemera. These stand-ins do not talk back but obligingly lend themselves to passive and noncontradictory analysis. Simplifications of this sort may be done in the name of dispassionate reason, as with uncritical positivism, as ungrounded literary or cultural criticism, or as ameliorative moral projects such as soul saving ethnography of the mission station or soup kitchen, or as urged by so-called standpoint epistemologies. In all cases, the yeasty, ambivalent, amorphous experience of social life is set straight, held at arm's length, narrowed and sanitized in the names of procedural or political propriety.

Postmodernism's strength is the encouragement it lends to varieties of aesthetic and critical writings that together may add usefully to the social sciences. But merely claiming postmodern allegiance is not enough. We must rid ourselves at once of the intellectual sclerosis of positivism and works in the name of science that are merely methodical in other ways, that are without art or craft. To borrow from Feyerabend, we do not wish to exchange the professional incompetence of modernism for an equally inconsequential incompetent professionalism, a self-satisfied postmodernism without human roots.

EPILOGUE

A few weeks before sending this article to press I visited Huntley again. This time, the advantage of surprise was mine.

I used a different magazine to start our conversation. After neutral greetings, I gave him a copy of the lead article from the May 1995 issue of *Atlantic Monthly*—"The Diversity Myth: America's Leading Export." There was a message in this gift: I was not an ordinary coin shop customer. He could tell I knew something about him, but what? Was I friend or foe? (I was not sure either.) The twelve years had dimmed recollections for both of us. As Huntley leafed through the article, he probed, "Uh, do you live around here? Have we talked before?" I affirmed both questions with a few details and, like the proud parent I am, I showed him a picture of my blond, blue-eyed, smiling three-and-a-half-year-old daughter. That was enough. Friend.

Huntley became amiable and, as before, instructive. The years had done nothing to diminish his eclectic literacy and eccentric zeal. He was soon launched into an impromptu lecture full of new facts and familiar themes. He warned again of the Franz Boaz legacy and the "spreading tentacles" of egalitarianism. He spoke of Shakespeare's disdainful view of Blacks, of the "abhorred union" of Othello and Desdemona, the ineptness of Portia's one Black suitor in *The Merchant of Venice*, and the villainy of Aaron from *Titus Andronicus*, the Bard's "Devil incarnate." Next, Huntley turned to the Rockefeller dynasty, starting with the patriarch, John D.'s father, "a bigamist and a charlatan," who traveled the byways of Pennsylvania and the East "fathering at least ten illegitimate children" and peddling raw petroleum as an elixir. The Rockefellers' interest in oil had its roots in this crude patent medicine, he told me, for "this stuff was just bubbling out of the ground . . . and nobody knew what to do with it . . . so they developed pharmaceuticals" and as a result "ninety percent of pharmaceuticals are petro based." He spoke of recent scholarship, of Edward O. Wilson's work in sociobiology, of *The Bell Curve*, even a recent article from *Society*, "The Seeds of Racial Explosion," by University of Southern California (USC) economics professor, Timur Kuran. Huntley continued for ten more minutes. As before, he seemed enthused by his topics and pleased with his audience.

There was something different about this visit. I felt more at ease, in control. Huntley's ideas were indeed strange but not unprecedented. In the past dozen years I had heard the likes of them more than once. I was on familiar ground. Half-attention was all I needed to follow his arguments. I listened but also looked around the shop. Things had changed. A few years ago Huntley had moved his business downtown, across from the county courthouse where he could "keep an eye on things better." (The police chief later told me this surveillance works both ways.) Huntley was visibly older. His hair was still thick but near white. His waist was still thin but now so were his arms. The furniture had been rearranged. Huntley's desk—and his gun—were now in an adjacent room, nearly out of sight, nearly out of reach. We talked standing at the counter this time and I noticed I was three or four inches taller than he. Judging from the posters on the walls and windows, Huntley is more active in community theater and music than politics these days.

It was late afternoon, time to close. The comforting institutional shadow of the courthouse crept up Huntley's storefront. Bidding him goodbye I walked toward home, calm and satisfied with the results of the day. All had gone well. I found new data and no real danger. Yet my composure was not full depth. It never is these days. Then and now I wonder about the other ways this interview might have gone. I have been face-to-face with a good deal of extremism in the past twelve years, at meetings of the Klan and the Aryan Nations, at clandestine training camps and public conferences, among antigovernment militias, messianic zealots, and would-be revolutionaries. Not all these encounters have been so serene or civil, but I am always lucky, I tell myself. Just be prepared and know what to expect—that's how to stay out of trouble. I was not surprised by Alan Berg's assassination or the Oklahoma City bombing. I have heard plans for similar events being discussed often enough. Mostly talk, I tell myself. Yet, lately I have grown uncertain, apprehensive. I wonder, have I done everything necessary for the next interview? New tape recorders run silently. And with a permit, I could carry a gun.

NOTES

1. In keeping with grounded theory methods, Kathy Charmaz developed these guidelines before reviewing the literature on fiction writing. They derive from her earlier work but strikingly echo strategies of fiction writers (see esp. Carroll 1990; Frank and Wall 1994; Giovannoni 1972; Hale 1972; Noble 1994).

2. Huntley is a pseudonym. All quoted material in this discussion of the first interview, unless otherwise identified, is from transcripts.

3. The sources of material attributed to Huntley in this paragraph are intentionally omitted. They derive from public documents that identify Huntley by his proper name.

4. The source of these comments is again deliberately omitted.

5. Style means the presence of the writer in the writing and reflects how the writer conveys his or her thoughts (see Barzun 1975; Lambuth 1976; Strunk and White 1959). Tone, an element of style, reveals ambiance and the writer's attitude (see Packer and Timpane 1986).

6. Postmodernist writers can err by attending too much to themselves and too little to their collective stories. Gary Provost (1980) contends that writer intrusion only works when the writer was a participant or possesses special expertise. Writer intrusion differs from voice. The writer's voice lets the reader imagine him or her speaking. A writer creates a voice by drawing on his or her perspective, vision, sentiments, and humor.

7. Richard's narrative distance fits his story. Narrative distance is not merely a ploy to gain or to claim objectivity—rather, the relative degree of narrative distance should fit the story and, in turn, fit the experience. Richard's telling of the tale brings him deep into the narrative. My involvement is in putting the analysis together, not in the story itself, and fosters much more narrative distance (see Atkinson 1992; Lofland and Lofland 1994, Richardson 1990; Wolcott 1990).

8. Michael Agar (1990) argues that creative nonfiction techniques in ethnographic writing can lead to reshaping the research process to fit the text. That is possible if we put the story before the experienced process. If, however, we put the process first, with all its nebulousness and stickiness, then we can use fiction techniques for purposes consistent with Agar.

REFERENCES

Agar, M. 1990. Text and fieldwork: Exploring the excluded middle. *Journal of Contemporary Ethnography* 19:73-88.

Atkinson, P. 1992. *Understanding ethnographic texts.* Newbury Park, CA: Sage.

Barzun, J. 1975. *Simple & direct: A rhetoric for writers.* New York: Harper & Row.

Bickham, J. 1994. *Setting: How to create and sustain a sharp sense of time and place in your fiction.* Cincinnati, OH: Writer's Digest Books.

Carroll, D. 1990. *A manual of writer's tricks.* New York: Paragon House.

Charmaz, K. 1991. *Good days, bad days: The self in chronic illness and time.* New Brunswick, NJ: Rutgers University Press.

———. 1995. Between positivism and postmodernism: Implications for method. In *Studies in symbolic interaction*, edited by Norman K. Denzin, vol. 17, 43-72. Greenwich, CT: JAI.

Cheney, T. 1983. *Getting the words right: How to rewrite, edit & revise.* Cincinnati, OH: Writer's Digest Books.
Denzin, N. 1986a. *The alcoholic self.* Newbury Park, CA: Sage.
———. 1986b. *The recovering self.* Newbury Park, CA: Sage.
———. 1989. *Interpretive interactionism.* Newbury Park, CA: Sage.
Ellis, C. 1995. *Final negotiations: A story of love, loss, and chronic illness.* Philadelphia: Temple University Press.
Frank, T., and D. Wall. 1994. *Finding your writer's voice: A guide to creative fiction.* New York: St. Martin's.
Giovannoni, J. 1972. 8 steps to professional writing. In *The creative writer,* edited by Aron Mathieu, 31-37. Cincinnati, OH: Writer's Digest Books.
Golightly, B. 1970. The use of dialogue. In *The writer's digest handbook of short story writing,* edited by Frank A. Dickson and Sandra Smythe, 58-65. Cincinnati, OH: Writer's Digest Books.
Hale, N. 1972. A note on feeling. In *The creative writer,* edited by Aron Mathieu, 118-24. Cincinnati, OH: Writer's Digest Books.
Hubbard, F. A. 1988. *How writing works.* New York: St. Martin's.
James, P. D. 1989. One clue at a time. In *The writer's handbook,* edited by Sylvia K. Burack, 204-7. Boston: The Writer.
Kogelschatz. 1983. Survival treasure chest: Today's pieces of eight make sterling investment. *Survive,* pp. 38-39, 41.
Krieger, S. 1984. Fiction and social science. In *Studies in symbolic interaction,* edited by Norman K. Denzin, vol. 5, 269-87. Greenwich, CT: JAI.
Lambuth, D. 1976. *The golden book on writing.* New York: Penguin Books.
Lofland, J., and L. H. Lofland. 1994. *Analyzing social settings.* 3d ed. Belmont, CA: Wadsworth.
Mitchell, R. 1983. *Mountain experience: The psychology and sociology of adventure.* Chicago: University of Chicago Press.
———. 1993. *Secrecy and fieldwork.* Newbury Park, CA: Sage.
Nash, W. 1989. *Rhetoric: The wit of persuasion.* Oxford: Blackwell.
Noble, W. 1994. *Conflict, action & suspense: How to pull readers in and carry them along with dramatic, powerful story-telling.* Cincinnati, OH: Writer's Digest Books.
Oates, J. 1970. Preface, "The nature of short fiction; or the nature of my short fiction." In *The writer's digest handbook of short story writing,* edited by Frank A. Dickson and Sandra Smythe, xi-xviii. Cincinnati, OH: Writer's Digest Books.
Packer, N., and J. Timpane. 1986. *Writing worth reading.* New York: St. Martin's.
Provost, G. 1980. *Make every word count.* Cincinnati, OH: Writer's Digest Books.
Prus, R. 1995. *Symbolic interaction and ethnographic research: Intersubjectivity and the study of human lived experience.* Albany: State University of New York Press.
Richardson, L. 1990. *Writing strategies: Reaching diverse audiences.* Newbury Park, CA: Sage.
Sartre, J. [1953] 1966. *Being and nothingness: An essay on phenomenological ontology.* Translated by Hazel Barnes. New York: Washington Square Press.
Solomon, Robert C. 1988. *Continental philosophy since 1750: The rise and fall of the self.* Oxford: Oxford University Press.
Strunk, W., and E. White. 1959. *The elements of style.* New York: Macmillan.
Wolcott, H. 1990. *Writing up qualitative research.* Newbury Park, CA: Sage.
Wright, L. R. 1989. How to keep the reader turning the pages. In *The writer's handbook,* edited by Sylvia K. Burack, 238-41. Boston: The Writer.

Yolen, J. 1989. Story-telling: The oldest and newest art. In *The writer's handbook*, edited by Sylvia K. Burack, 489-93. Boston: The Writer.

RICHARD G. MITCHELL, Jr. is Associate Professor of Sociology at Oregon State University. His research interests are philosophies of fieldwork and millennial social movements, including survivalists, supremacists, and retreatists.

KATHY CHARMAZ is Professor and Chair of the sociology department at Sonoma State University. She was the recipient of the 1992 Charles Horton Cooley Award from the Society for the Study of Symbolic Interaction and the 1992 Distinguished Scholarship Award from the Pacific Sociological Association for her book *Good Days, Bad Days: The Self in Chronic Illness and Time*. For nine years, she was enrolled in the Alternative English Major at Sonoma State University and for six years taught courses on writing in the School of Humanities.

Chapter Twelve

Writing a Case Study

Case studies are not merely class assignments. They are a useful empirical and analytical approach to understanding individuals or groups, written to help us understand larger social processes. Interviewing a social worker for a day, following that person around at work, and asking questions about his or her job can help you understand a lot about the nature of working for a social service agency. Talking to a homeless man over a cup of coffee can tell you a lot about the ways that homeless people experience stigma and shame. Meeting with a group of construction workers at a work site may reveal to you important things about how people in physically demanding work think about the possibility of workplace injuries. Sometimes a case study is a data point, to be placed next to other case studies. Other times, it is a study in and of itself.

A few years ago the leaders of a high-tech company asked me to help them understand the internal culture of their organization, particularly around the topic of safety. My research team was not asked to compare multiple companies, just to study this one case. So as I worked with them on this assessment, I treated it as a case study of a place that had a reputation for focusing attention on safety; hence, it was a potentially unusual case. By thinking about what this was a case of (e.g., an organization with unusually safety-conscious leadership), I was able to draw attention to what we could learn from this one special case. We learned that the company's particular ways of addressing safety were producing unintended lethargy and smugness about safety among the workers, thus reducing their attentiveness to safety. This was not good news for the company.

Our study was not an ethnography, meaning that we were not permitted to spend weeks wandering around the factory or working there. Nor was it a systematic survey of workers. But we were able to produce for the company's leadership a profile of how the way they were training workers about safety was having unintended and undesirable consequences (Edwards and Jabs 2009).

In our safety study, we were pretty sure that the company was not a representative one that resembled all kinds of companies. But we could treat it as a very good example of a company that was doing as much as one could expect to promote safety. So in this sense, the company was studied because it was unusual rather than because it was representative of most other organizations. Going into the study, we did not know if this case would really teach us anything, but we discovered that this kind of case study permitted us to apply our sociological imagination to a new situation and propose some interesting hypotheses for further research.

What to Focus on, What to Report

Sometimes in your sociology classes you'll be asked to take a close look at an individual, social group, or unit and to identify in case study format the same kinds of things we looked at in our safety culture study, namely,

- relationships between individuals (or groups) and others;
- problems that they are facing and how they are addressing them;
- what individuals or groups consider good or bad, important or unimportant;
- their normal routine activities and deviations from those activities; and
- how social structural influences (circumstances, rules, policies, powerful others, etc.) seem to shape what people experience.

When writing a case study, you may be tempted to discuss every individual, conversation, or relationship between people. But after considering all your observations and conversations, you will have to consolidate your ideas and illustrate them with what may seem like relatively few illustrations, examples, quotes, stories, and so forth. In the end, you'll need to organize your case study around sociological concepts, processes, and patterns rather than merely treat these bulleted points as a checklist. The trick is providing enough details, quotations, and observations to illustrate your points while keeping it conceptually organized.

Tellis (1997) points out that case studies can incorporate what people say, what they do, and what they read, write down, or work with. In our safety study, we were permitted to talk to some of the workers about their ideas about safety, but more important, we reported on how different groups of workers interacted with each other or how they said they related to their supervisors. We focused on what they thought others were thinking and on their interpretations of events and company policies. We looked at safety program brochures and black-and-yellow hazard tape on water coolers. By the end of our long case study, we wanted the readers to get a sense of what it was like at this particular company, and we did so through mixing description (so they could see and hear what was there) and interpretation (so they could understand what we thought these things meant).

Some Advice about Constructing Your Case Study

As with any writing, you should always ask yourself who your audience is. For a class assignment, your professor will certainly be the primary audience, but you would do well to imagine at least a larger audience of intelligent students like yourself who may not be familiar with the setting, the research questions, or the literature that is relevant to your case. Whom you imagine as the audience will shape the level of detail you include and the ways you provide interpretation to that detail. Here are some suggestions:

1. *Opening the case:* It is most effective to begin with a description of the biggest issue being addressed by the individual or group. The opening sets the scene for subsequent challenges, conflicts, and responses. For example, if you had a simple case study of one social worker, you might start with, "Social workers struggle with the tension of balancing their obedience to social program rules and advocating for clients. My daylong interview with a county social worker, hereafter referred to as Jane, illustrates how she manages to navigate these competing demands." Notice that now the reader knows the big issue or problem (the balancing act), where you got your information (your day with Jane), and what we will now learn (how Jane deals with this balancing act).

2. *Telling the reader enough detail to ground the story in a real setting:* The reader should be able to imagine you on the scene because you provide enough detail to situate the case. For example, you might say,

> Jane is expected to work from 8:00 to 5:00, spending about half of her time in the office and half of her time driving around the county to visit people. On the day I visited, however, she worked until 8:00 p.m. In her beat-up, county-owned white Ford Focus, we were returning from a remote rural home after sunset when she told me, "This is not unusual to be working so late during this time of the month." I asked about her social worker colleagues, and she said that pretty much all of them put in the overtime like she does.

In short, tell the story of your being there, but organize your story around the main themes you are trying to explore with your case study. (See chapter 11 in this handbook about telling stories with ethnographic data.)

3. *Connecting your case to larger processes:* Because your case study is meant to illuminate or illustrate larger social processes, look for ways to keep linking your case back to the items you have read about in the sociological literature. For example, you might say, "Jane's observation about how she finds loopholes in the policies to provide extra financial help to her clients illustrates Lipsky's concept of street-level bureaucracy wherein social workers and other public employees face competing expectations. Lipsky argues that . . . , and Jane's statement about . . . illustrates this perfectly." Notice that you have the opportunity to teach the reader what you have learned about how processes observed in this case relate to what others have already said. If you are writing this for a class project, you may be asked to link your observations to course materials, but if not, connect it to the research literature. This implies, of course, that you will need to do some reading, one hopes ahead of time, to help you know what that literature is and how your observations are linked to it.

4. *Wrapping up:* The reader should be able to see which new questions, hypotheses, and/or conclusions emerge from the case study. A quick summation at the end that captures the main points made in the case study will serve as evidence that you have a clear picture of what this case study can teach us and where it should lead us next.

References

Edwards, Mark and Lorelle Jabs. 2009. "When Safety Culture Backfires: Unintended Consequences of Half-shared Governance in a High Tech Workplace." *Social Science Journal* 46:707–23.

Tellis, Winston. 1997. "Introduction to Case Study." *Qualitative Report* 3(2). Retrieved January 7, 2011 (http://www.nova.edu/sss/QR/QR3–2/tellis1.html).

Chapter Thirteen

The Internship Journal

Gary Tiedeman

Internships are intended to provide students with opportunities to enhance their education through activities performed outside the classroom (paid or voluntary) within a work setting in the real world. The setting will ordinarily be a social service agency, nonprofit organization, or some sort of business establishment. This chapter speaks to one central component of the typical internship experience: the student journal.

As a starting point, an important fact should be kept in mind: Usually, internship credits are awarded not simply for faithfully participating in the work setting but instead for demonstrating creative, analytic, sociological thought and intellectual growth as a consequence of day-to-day work activity. Without this criterion, the academic department would be placed in the awkward position of granting academic credit for a potentially nonacademic experience. Therefore, the faculty supervisor must have some means, written and/or oral, of judging the student's academic achievement. Those means might include (depending on specific arrangements made between each student and his or her faculty supervisor) a research paper, reviews of relevant books or journal articles, periodic face-to-face meetings, or perhaps most common, an internship journal.

What should that journal be? How should it be organized and conveyed? What is appropriate content for it? The first trick to becoming a successful

---

Author's Note: My colleague Gary Tiedeman, now retired, supervised hundreds of student internships during his long career. Here he provides some excellent advice about writing an internship journal.

academic journal writer is to always think of yourself as a storyteller. Your particular story is telling about the sociological background, social influences, social structure, and social processes that characterize the setting in which you are working. Do not take anything for granted, and do not assume that the reader can grasp the subtleties, relevance, and context of the situation without your describing them. Your job is to convey the substance and significance of your topic with the same ample detail that you would appreciate if you were having the same story told to you. "Cinderella," although basically a tale about domestic abuse, about the startling transmutation of mice and pumpkins into horses and carriages, and about romantic dreams come true, is far more captivating in its familiar "once upon a time" form of delivery than as a case entry reading: "Destitute, delusional young female exhibited hallucinatory behavior today." You need not write a fairy tale, a novel, or even a short story depicting every incident at your work site, but take the time and effort to flesh out a few particularly interesting situations in some length and detail. Try it; you'll probably find that you enjoy it—and learn some things you hadn't thought about before!

It might be easier to visualize what the journal should be by first visualizing what it should not be. The journal should not be limited to descriptive summaries of what occurred on a given day, no matter how lengthy or detailed. This is where a great deal of confusion arises in terms of the discrepancy between what may come most readily or automatically for the student, on the one hand, and what the faculty supervisor wishes and expects, on the other. It is perfectly fine (in fact, it is proper) to begin each journal entry as a dated, diary-format summarization of what the student has experienced at the internship site. But the premium from the instructor's point of view is on what the student is learning and applying from those same experiences.

Some students prefer daily entries, but weekly or twice weekly entries may be acceptable depending on site circumstances. It is important not to let too much time elapse between entries, because important details can be quickly forgotten. Also, it is often during a rigidly scheduled daily (or nightly) write-up that the best insights, applications, and connections come to mind. Brief, general activity descriptions are normally ample; exhaustive detail should be avoided unless it serves a real purpose in making a particular point or in depicting the special, fine-tuned nuances of an unusual problem or situation. But more important than the descriptive details is the sense the student makes of it; this making-sense-of-it sort of discussion is the most vital part of the journal and probably the single greatest key to making the internship a valuable educational experience rather than just a work activity alone. Why did things happen the way they did? How does an incident relate to other aspects of the organization or to particular personnel or personnel functions? Is there a consistency or inconsistency between related incidents or situations? How might things have been handled differently, if at all? Does the real experience agree with or contradict what courses and textbooks have had to say about it? What have you learned today that helps make better sense of a confusing or frustrating occurrence of two weeks ago? How so? Et cetera, et cetera, et cetera. A properly compiled journal, in short, confirms that the student is truly learning and that his or her mind is alive, as opposed to simply moving through mechanical routines unthinkingly.

So let's review, from a slightly different perspective. What you should seek to create is an analytical journal, as distinct from a purely descriptive journal. A descriptive journal, which many students tend to write and submit, simply records the main events that occur day to day on the job. In extreme form, the reader/evaluator might find just a one-sentence or one-paragraph mention of a single noteworthy task, event, or activity from a given day on site. (Even at the purely descriptive level, there should be a more lengthy and more detailed account of a broader span of the day's experience.) More important, however, the point is that the academic evaluator needs to be able to see (i.e., to read) what learning experience(s) took place. Otherwise, the evaluator finds him- or herself in the awkward and impossible position of attempting to evaluate academic performance on the basis of purely physical behaviors. A descriptive journal tells nothing about what is being learned to justify the granting of the academic credit that the student seeks.

An analytical journal, on the other hand, begins with the same reporting of events but then intersperses sections of commentary and discussion, which show the reader that a sociological perspective is being applied. Let's say that the setting is a social agency of some sort and the student is reporting an interesting case contact that occurred on a given day. The rudimentary, purely descriptive journal entry might simply state, "Dealt with an interesting case." Period! The more expansive entry (of the type desired by the evaluator) first gives added detail about the nature of the case and what makes it particularly interesting. (An important side note is to always use pseudonyms, not actual names, when referring to any client or customer.) Was it the issue itself that made the case interesting and worthy of added thought and discussion? If so, how? Why? Was it the people involved? If so, why? How? What aspects or characteristics were most pertinent? Is it a combination of issue and participants that creates the interest? If so, describe the interaction of the two.

Next, now that you have fleshed out the descriptive basics, go on to analysis, implications, and/or applications. Why do you think things happened the way they did? What is it about the organization; or about the organization's policies, rules, regulations, assumptions, or standardized approaches; or about the people involved that provides an accounting for the incident? Which concepts, perspectives, or theories from your academic course work have a possible bearing? Identify them, and talk about how they fit—or if appropriate, about how they fail to explain what they're supposed to be able to explain. In other words, does the classroom and textbook theory match what you see as being the reality? If not, how does the academic material need to be adjusted or updated in terms of the insufficiencies you have discovered? For example, did the researchers who formulated a particular theory or concluded their article with a set of statistically significant findings fail to note a variable that you consider all-important in your setting? What would you call that variable? How would you describe it? How would you define it operationally and measure it?

Three Kinds of Internship Journal Entries

Very Weak	*Better*	*Well Done*
Today I answered six phone calls from potential clients and received eight visitors to the office, all asking about how to apply for food stamps and unemployment.	I answered six phone calls, all inquiries about unemployment and food stamps, all from women who said they were single with kids. Another eight people, five who were single mothers, also came to the office asking about the same thing. All of the people I dealt with had some form of misinformation about how to apply or about what might make them eligible.	Based on the 6 phone calls and 8 visits from people asking about unemployment and food stamps, it was clear that people are hearing false information from others sources. Perhaps these are rumors spread by others who have experience, or perhaps this information comes from other social service agencies and nonprofit groups. If other service groups are handing out incorrect help to the poorest people in our community, this raises an important question about how the experience of being poor is made worse by the very people trying to help. Of the 14 people I talked with, 11 were single mothers. So if other groups are making it harder on these poor families, it has real implications for how children are affected by the lack of clear organization among the various service agencies. Lacking clear information seems to be an indicator of powerlessness.

Three Kinds of Internship Journal Entries *(Continued)*

Why?	*Why?*	*Why?*
Weak because it merely describes in a cursory way the activities of the day.	Stronger not just because it is longer but because it begins to explore some patterns, linking family status to inquiries for help and observing something unexpected among the people encountered.	Shows some analytical effort, providing details (numbers, how informed people are, gender, and family status), but also linking these issues and reflecting on larger structural influences, unintended consequences, implications, and so forth. In other words, shows sociological imagination.

As for implications and applications, show some thinking (in print) about such topics as the effect of changes or difficulties in one aspect of the organization on other aspects of the organization. This is an effort to demonstrate your understanding and appreciation of the total operation as a social system rather than as a collection of independent features (e.g., Does low salary cause low morale, which in turn causes low commitment and shoddy work? Are parallel situations handled so differently by various staff members that organizational inconsistency and confusion results? Is training adequate for performance needs? Is there a two-way flow of communication up and down the organizational hierarchy?).

Similarly, experiment with suggestions that you identify for organizational modifications (in either structure or process or both), for new directions the organization might take—or might find itself forced to take against its real wishes—or current activities, topics, or functions that the organization might consider eliminating because of new priorities, expectations, or focuses. These might very well be suggestions or insights that you would not necessarily share with organization supervisors, but they can be very helpful in giving the academic evaluator a sense of your trajectory of learning and intellectual growth.

These suggestions outline several of the ways available to you to add depth and substance to your journal content. You won't be expected to cover everything described, nor should you attempt to go into equivalent depth with every single journal entry. Also, keep in mind that you shouldn't limit yourself to the types of questions and examples illustrated here. Each internship setting has unique features that allow unique observations and interpretations. What you read here is meant as a guide, not a mandatory standard. The social universe is highly varied and ever changing, and part of your job is to discover how to best adapt the academic learning element of the internship to the features of your particular setting. With these guiding principles and suggestions in mind, proceed and enjoy! It is practically guaranteed that the end result will be to make your internship experience a more meaningful and valued one for you.

Section IV

Other Sociology Writing Tasks

Chapter Fourteen

Revisiting Literature Reviews

Applied Sociology Research Projects

Denise Lach

Determining research focus is somewhat different for applied research than for more basic research efforts. In applied settings, the social scientist does not set the agenda but acts in partnership with a variety of client groups. This is particularly true for the task of defining the problem and deciding on the research questions. Therefore, in many applied studies, the literature review provides background resources rather than the essential starting point for research designs.

While critical to strong applied research, a good understanding of what is already known or established does not have the absolutely central role that it does in fundamental, discipline-developing research. In addition to reviewing published results of research, you may need to get relevant

Author's Note: My colleague Denise Lach has spent years successfully using her sociological perspective to help solve interesting applied social science questions for public agencies. Because she often is writing literature reviews for applied research projects (some quantitative, some qualitative), convincing funding agencies to take seriously the questions they really need to pursue, I asked her to describe what she thinks about and actually does when she is writing a literature review. You will see that much of her advice resonates with what I have indicated earlier, but she also has some unique suggestions about this kind of writing.

background information from persons who have done related work, review program-specific documents and data, and talk with individuals who have participated in the development or implementation of activities. Unfortunately, for many applied topics, the literature and other data tend to be somewhat inaccessible and fragmentary. People refer to this body of work as "gray literature," and it is available primarily from the sponsoring organization (e.g., agency, company, or individual).

The applied literature review serves multiple purposes, including

- setting your problem in a context in a way that convinces readers of your paper that it is an important issue,
- describing research questions that you are pursuing in this effort,
- examining the literature and providing a comprehensive survey,
- identifying important variables, and
- describing the model you are using to explain the relationships among the variables.

Proposals for applied research require a great deal of upfront discussion that goes beyond classic literature reviews. It is crucial to clarify and refine the scope of research prior to investing significant time and effort. Scoping activities are necessary to provide background about the problem, clarify the specific interests of the research sponsors, and identify any constraints or boundaries that are likely to affect the research. Scoping activities include the following:

- discussions with clients to obtain the clearest possible picture of their concerns;
- review of the relevant literature, including research reports, transcripts of legislative hearings, program descriptions, administrative reports, agency statistics, media articles, and policy position papers—this literature should provide some historical context as well as clues for approaches to the research question or existing data;
- gathering together of current information from experts on the issue and other major parties to understand the current context and possible areas of disagreement or debate; and
- information-gathering visits to the sites of the program or problem to obtain a real-world sense of the context and to talk with people who are actively involved in the issue.

Much of the information collected through this scoping process, while not typical of basic research, will need to be included in the literature review of an applied research report.

Literature Review Tips for Applied Research Reports and Proposals

A good literature review for an applied social science research project should

- define a problem and its scope,
- ask one or more research questions,
- examine the literature,
- report a survey of the literature (not an exhaustive report),
- identify the important variables that affect the problem,
- develop a model to explain the interaction of the variables, and
- propose the usefulness of the model to the field of study (e.g., sociology).

Typical writing problems in literature reviews include the following:

Unclear scientific and practical purposes of the paper.

Solution: Try writing at a more conceptually abstract level when discussing the background to this problem or issue. This will make you more able to generalize and to link your work to other existing research on the topic. *Be cognizant of the differences between generality (comprehensive, wide applicability) and vagueness (imprecision, shallowness).*

Items and ideas are misplaced in the presentation.

Solution: Writers naturally, understandably, and frequently (but wrongly) place items in a paper in the order in which they think of them rather than where the logic of the paper requires them to be located. *Keep clearly and directly to the subject of your paper's (sub)headings.*

Separate sections seem to lack purpose with respect to the paper as a whole.

Solution: Do more than go through the motions in each section; focus on and tell your readers what each section accomplishes for the paper. *Make sure that you distinguish between writing a review as a summary (which is incomplete and unacceptable) and a review that serves to integrate or critique (which is what is expected and required of this type of review).*

The following are some specific tips for good writing:

- Write in the active voice. Use a grammar-checking software package if necessary to find and remove all passive voice.
- Before writing, develop a clear purpose statement for each paper. If you cannot finish the sentence, "The purpose of this paper is . . . ," then you are not ready to write.

- After you are done, create a title that clearly states what the paper does or whom it serves.
- Use subheadings liberally. These guide the reader through the paper.
- Ensure that each section transitions into the next section.

An Extended Example: What's Really Happening When I Write a Literature Review?

To get started, I usually do a search on my key variables (such as homelessness, minimum wage, industrial forests, organizational decision making) to see what I find and also to identify what appear to be other key variables that I may not have been smart enough to identify up front. This search usually happens very quickly as I scan titles, abstracts, and so forth. I also try searching for different combinations of variables, similar words, or synonyms to see if I can find anything close to my problem or topic. Then I use multiple strategies for extending my search, for example, looking at the bibliographies of articles or books on my topic (or closely related topics) because they are a good source for finding other references.

It is important to learn how to scan articles or books very quickly. If you read every word, you'll be ready for retirement before your literature review is ever finished! One way to keep yourself reading/scanning quickly is to look at the abstract and introduction (although I know that many people skip the introduction because they think that it repeats the abstract too much). I like the introduction because I can usually find the problem statement clearly stated. If the abstract and/or introduction looks good, I retrieve the document. I know that I'm not finished after this first pass, but this is usually a good place to stop before I'm overwhelmed or bogged down.

After I finish checking out books and requesting books that are already checked out (!) as well as downloading or photocopying articles I think will be useful, I begin reading. As I read, I highlight and write down important ideas or questions that should remind me of the major arguments and points in the text. And I always write a short note to myself about what purpose the article or book will serve (how it will help the discussion about variable X or relationship Y, e.g.). Also, as I'm reading, I make notes about other sources of information that might be useful. This is not an endless task, but it does take some time. When I'm finished with the first bunch of materials, I go back to the library or the Internet with my list of new sources that was generated in my first pass through the material. Some literature reviews will be massive (as big as a book), and others will be minimal. It depends in large measure on the timeline under which you are working. You have to decide at which point you're satisfied that you have the information that you need to begin writing. You can always go back to the library if you discover you need more information.

Once I have a number of texts, I start sorting the texts into logical groupings. (Don't wait too long to do this step!) I look at the literature to see where there is agreement or disagreement on the topic. I'm trying to see whether there are holes in the literature that I've collected (and I need to go back for more) or holes in the research (areas that I find aren't very well covered). Also, I'm trying to keep my mind open to any new research questions that come up as I read.

Then I begin writing—always the hard part—within these groupings. I start with a summary of findings in one of the groupings and then review specific articles or lines of argument in some detail. Most literature reviews provide pretty extensive reviews of one or two key pieces of research that exemplify what's going on in the field and then summarize the rest. I also include my own conclusions at the end of each of these sections about what's missing, what's messy, and so forth. At the end of the literature review (you'll get there!), I summarize by giving a very general overview of the literature and discussing the problems and opportunities for my own research, indicating how my project speaks to this literature.

Here are a couple of paragraphs from a literature review from a funded proposal. Look at how we use the literature to frame our problem and suggest that existing research and methods don't work. I've inserted some editorial comments in brackets at critical points.

> The most popular social science model for decision making is the rational choice perspective. This model suggests that resource management choices are (or at lease strive to be) based on a search for information, followed by comparisons and weighing of information, leading to selection of the best alternative. *[This model is pretty much common knowledge in the community of people who will read this proposal, so we don't provide a citation, although we could, and maybe even should, do so.]* The rational choice approach suggests that ENSO *[El Nino Southern Oscillation]* forecast information will be readily incorporated in decision making (Beyer and Trice 1982). Although it is based on individualistic assumptions of utility maximization which render it unsuitable for collective decision making (Arrow 1951) *[I have to admit that we are showing off a bit here by using this classic economic model to make our argument.]*, the rational choice model is usually assumed to be applicable at the level of organizational decision making, either by breaking down organizational processes to individual decision points, or by treating each organization as if it were a unitary individual—a person writ large (Jaeger et al. 1998). *[This paragraph sets the baseline for our argument and uses other sources pretty sparingly; however, we go in for the kill in the next paragraph by citing everything and everyone that contradicts these assumptions.]*

However, studies of actual decision making in public and private sector organizations indicate that the rational choice model may not be the appropriate one for institutional decision making (Douglas 1986). *[Notice how we start this paragraph by contrasting with the previous paragraph, beginning with "however."]* In particular, the knowledge use literature suggests that information is not very well used in organizational decision making (Gurvitch 1972; Argyris 1987; Argyris and Schon 1978; Holzner and Fisher 1979; Caplan 1983; Dunn 1983; Averich 1987). *[Usually, for empirical results, you want to stick with the most recent sources. But sometimes it's possible to build up a history of research that supports your argument.]* Empirical studies show that institutional decision makers have a generally positive attitude towards the use of scientific information in decision making, but rarely act upon such information directly (Starling 1979; Weiss and Bucuvalas 1980; Whiteman 1985; House and Shull 1988). *[We then go on to discuss a single article in greater detail.]*

This gives you some idea of how I use the literature to frame and support my problem, conceptual framework, and ultimately, research methods. Indeed, some literature reviews can run to 50 pages. But in most papers or articles, the literature review will be much shorter. That in itself is a problem because you need to think hard about what's important and how to support those things you think are important through the available literature. It's not enough to just throw in every bit of literature that you come across without thinking about how it supports your purposes in the paper.

Chapter Fifteen

Writing a Book Review

Many college students, when asked to write book reviews, recall the many book reports they had to complete in elementary, middle, and high school. But a book review in sociology is not the same as a precollege book report. We do not organize our thoughts around such things as character, plot, or setting, and we are not merely trying to show our teacher that we read the book. The purpose of a book review is to help fellow social science students and colleagues understand if and how the book accomplishes something useful and how that accomplishment was achieved. When this is done well, some people will say, "That sounds like a book I need to read," and others will say, "Thanks for warning me off from that loser."

Through your college training, you are learning to think sociologically. Therefore, think about what like-minded people would want to know. That is, keep your audience in mind, informed by what you know you have in common with them. Your sociological training tells you that theory and methods matter, that the discipline has some order and structure to it, and that our work often has real-life implications. So consider how your book review will communicate to fellow sociologists how the author builds her or his argument, how the book relates to sociological theory (or at least which compelling questions in the discipline it tries to answer), how the author accomplishes the work, and what the implications of the book could be. Because the book is meant to speak to a larger sociological literature,

---

Author's Note: This chapter began as a chapter by Lori Cramer in the Oregon State University sociology department's original handbook and then was further developed by Mark Edwards for this publication.

readers usually want to know what you think this work contributes to that literature. Does it reaffirm things we already knew or provide new, unexpected results? Does it radically reorient our thinking to a topic, solve a dispute, or illustrate something we suspected but about which we were not sure? These are the kinds of issues that book reviews are meant to address.

Hartley (2006) surveyed professors in the social sciences, the biological sciences, and the humanities to ask about their experiences writing and reading book reviews. He found that there is considerable agreement across the disciplines about the important central elements of book reviews. First, professors report that book reviews become "dreadful" when they spend too much time summarizing the content and not critiquing it, when the author fails to discuss the book's argument and worth, and when the book reviews are "too short, long, terse, shallow, pedestrian, self-serving, bitchy, negative, sarcastic, etc." (Hartley 2006:1200). Ouch. Let's not write those kinds of book reviews.

The professors agree that what they value in good book reviews are at least these kinds of elements:

- a straightforward overview of what the book is about,
- a critique of the argument of the book,
- an evaluation of the book's academic credibility,
- a comparison with other works in the field, and
- an assessment of the book's usefulness for its intended audience.

That laundry list is not an outline, but you can see that readers of book reviews will be looking for these kinds of things.

So how should you organize a book review to achieve these goals? First, plan an outline not unlike the dozens of essays you may have written to this point in your schooling. That is, arrange paragraphs so that they introduce the book, discuss the issues identified by those survey participants in the previous paragraph, and end with a final articulation of your recommendation about the success and failures of the book. In other words, your review should have a beginning, middle, and end. Second, rely on your other writing skills for the line-by-line construction of your paragraphs. All that you have learned over the years about writing effective paragraphs applies here—internal to each paragraph, use opening thesis statements and then supporting material to make the paragraphs internally coherent. It is often useful to rely on signpost language like we just used in this paragraph, using *first, second,* and so forth to structure your argument and defense.

So at the risk of turning this into a recipe, here are some guidelines for writing book reviews that can help you structure, evaluate, and edit your own book review.

1. Begin reviews by listing the facts of publication. That is, provide the same information that you would in a citation at the end of a research paper. (See chapter 5 for information about citing and referencing sources.) For example,

Johnson, Daniel M. and Rex R. Campbell. 1981. *Black Migration in America: A Social Demographic History.* Durham, NC: Duke University Press. 190 pp.

2. Include (a) a brief summary of the book's contents and central thesis; (b) your assessment and appraisal of the book's merits and shortcomings, for example, how it compares with other books on the same subject, whether its conclusions flow from the analysis, what the important findings or conclusions are, and whether there is anything new or different in them; and (c) some judgment as to its relevant audience, in particular, its usefulness to sociology. Achieving this may take several paragraphs.
3. Your review should minimize (a) anecdotal information about the author or the history of the book (e.g., "As friends since childhood, Johnson and Campbell collaborate on . . .") and (b) jargon and technical language (unless discussing issues where you can help the reader by defining and then using technical language). Remember that your audience is likely to comprise fellow social science students who are generalists, familiar with some sociological concepts but not the minutia and detail of its various subdisciplines. When writing your book review for a class, avoid writing as if only to your teacher, where you will be tempted to either assume he or she knows what you mean or to show your ability to use big words.
4. Your review should not (a) repeat the table of contents chapter by chapter or section by section, (b) go to great lengths to find something bad or good about the book or something that should have been included, or (c) state the obvious.
5. In the case of books that collect chapters by several different authors, place greatest emphasis on the quality of the book as a whole. Limit references to specific arguments or chapters except to award special praise or criticism or to illustrate the general points you wish to make.
6. Avoid quoting long passages from the book you are reviewing. Paraphrase when possible. Whenever you use a quote, give the page number of the quote.
7. Avoid using references and footnotes. If a quotation from another work is absolutely necessary, incorporate the reference into the text.

Remember that while these are guidelines, your reviewer (a teacher, a colleague, a journal, a graduate school admissions committee) may have slightly different expectations or would wish for you to emphasize slightly different things. But the basic structure and style should be similar across audiences and settings.

Finally, most important—read book reviews. The more you read them, the more you will get the feel for book reviews, seeing common structure and style elements that will help you write your book reviews with greater confidence and efficiency.

Reference

Hartley, James. 2006. "Reading and Writing Book Reviews across the Disciplines." *Journal of the American Society for Information Science and Technology* 57(9):1194–1207.

Chapter Sixteen

Tips on Writing Theory and Content Papers

Sheila Cordray

Two of the types of writing you will be asked to do as a sociology major or minor are theory papers and content papers. Let's begin by distinguishing between the two, although you may be asked fairly frequently to write papers that combine the two. A theory paper is one wherein you write about, or use, some sociological idea or concept to explain or understand some aspect of the social world. In a content paper you would focus on some particular aspect of the social world. You are probably most familiar with content papers under the label of "library research" papers. For these, you do not collect data yourself, but you use information collected by other people interested in the topic.

In a content paper, you might write about topics or content areas such as the family, the political institution, deviance, or natural resources. The focus is on the topic—American family, the Democratic Party, youth gangs, or sustainable forestry. In content papers you need to demonstrate your understanding of the topic. You will want to find out as much as you can about the topic area. This might mean collecting data from the census or other surveys, reviewing the literature to find articles from both scholarly and popular journals, and possibly doing some research of your own (e.g.,

Author's Note: My colleague Sheila Cordray, now retired, taught sociological theory for a couple decades. She has here distilled some of the most important things she communicated to many cohorts of her students about writing these kinds of papers.

interviewing gang members, visiting an industrial forest). You will be expected to describe the topic in sociological terms using concepts such as norms, values, roles, institutions, class, power, or deviance. Content papers are often assigned using prompts such as the following:

- Describe the current eating habits of the American family.
- Review the changes in the American political institution that have made the Democratic Party the minority party.
- Compare and contrast delinquent gangs and Greek organizations.
- Show the effects of the Endangered Species Act on the timber industry.

In a theory paper, the focus is on the sociological ideas that you use to understand what's going on in the social situations rather than on the topic itself. For example, you might use Weber's concept of rationalization to understand the changing American family, Mills's sociological imagination to account for Democratic politics, Durkheim's ideas about social solidarity to understand delinquent gangs, and Marx's work on commodities to look at what is going on in industrial forestry. In each one of these cases, the focus is on the theoretical concept or idea and how to use it as an explanatory tool. The topic often is taken for granted or is a given. You do not need to collect more information about it. You just need to answer the question posed from a theoretical perspective. Assignments would pose questions such as the following:

- Explain why pizza is America's most popular food using Weber's concept of rationalization.
- How does the sociological imagination help us understand why there are more Republicans in Congress than Democrats?
- Why do people join social organizations such as gangs or fraternities? Use Durkheim's concept of solidarity to answer.
- Use Marx's ideas about commodities to explain why it is difficult to do sustainable forestry in a capitalist system.

As an example, if you were asked to use the concept of rationalization to understand pizza consumption in the United States, you would not spend a lot of time collecting statistics about pizza consumption, the history of pizza, or the best places to get pizza. Rather, you would spend your time reading and thinking about rationalization and how the components of this concept (calculability, efficiency, predictability, and dehumanization) help you to understand why people in the United States eat a lot of pizza.

Clearly you could do both tasks in a single paper. You could describe a given social situation or problem from a sociological perspective and then use a theoretical concept or idea to understand or explain what is going on. Let's keep them separate for now, as you will frequently be asked to do one or the other. However, understanding the difference between the two tasks should help you write a combination paper as well.

Here's a summary of the differences between the two types of papers with some tips about how to proceed with your writing.

Element	*Content Paper*	*Theoretical Paper*
The focus of your paper	Focus on a social situation, problem, or topic.	Focus on the use of a theoretical concept or idea.
How you start your research	Identify sources of information: books, articles, websites.	Read about theoretical concepts and ideas in assigned reading or other sources.
The process of writing	Examine all the information you have collected about the topic, select a congenial and logical method of organizing the information, identify your organizing ideas (norms, values, roles, institutions, etc.) and any supporting ideas, and proceed to answer the questions posed in the assignment (adapted from Packer and Timpane 1989:43).	Begin with a brief discussion of the question so that the reader is familiar with the situation you are explaining, identify the concepts or ideas you will use to answer the question, define all concepts and explain all ideas, and use the concepts and ideas to answer the question posed in the assignment.
The process of revision	Make sure you have used information from a variety of sources and covered the issues posed by the question, identified the sociological ideas you have used to structure your paper, checked to see that all sources are appropriately cited and that your bibliography is	Use clear and specific conceptual definitions—ideas should be clearly explained with reference to texts or lectures—identify premises and make sure all assertions are supported, check the structure of the paper

(Continued)

(*Continued*)

Element	*Content Paper*	*Theoretical Paper*
	complete, used an outline of topic sentences from each paragraph to check organizational structure; and read carefully to detect any claims about the situation not supported by the data you have collected.	by making an outline of topic sentences, and review the logic of your argument to make sure that you have answered the question posed.

Reference

Packer, N. and J. Timpane. 1989. *Writing Worth Reading*. New York: St. Martin's.

Appendix: Word Use and Misuse

Gary Tiedeman

We all have trouble with certain words in the amazingly complex system known as the English language. What we have here is a partial catalogue of some of the most common stumbling blocks encountered in exams and papers written by social science students. The items are in alphabetical order rather than in any particular order of importance.

accept, except: To accept is to take willingly; to except is to skip or reject. Hence, both of the following are **correct:**

> "I am very pleased to accept your offer of employment for a base salary of $100,000 per year."
>
> "I like everything about him except his looks, his personality, and his behavior."

adverse, averse: Here's one that even college professors mix up. The main problem seems to be that most people aren't aware that there is such a word as *averse* (even though, interestingly enough, they might be perfectly at ease speaking about aversion therapy). In any case, *averse* describes a person's sensation of distaste or opposition to something:

> "I am averse to having my nipples pierced, thank you."

Adverse is a word used to refer to something that hinders or opposes progress, as in "adverse conditions."

> **Wrong,** but common: "I'm not adverse to that approach."

affect, effect: This is a tricky one, and even the best of writers sometimes have difficulty with it. Try to remember the difference by thinking of *effect* as consequence (a noun) and of *affect* as an action (a verb):

"His unusual sensitivity has had a profound effect on me."

"His unusual sensitivity affects my own view of the world."

To confuse things a bit further, the word *affect* can also be used in a psychological context to refer to a feeling or emotion. But this usage stands apart from the area of confusion cited above and usually presents no difficulty.

allowed, aloud: Easy. Just remember that the word *allow* never loses its *w* when it takes longer forms, such as *allowance, allowable*—or *allowed.* It always has to do with whether something is permitted. *Aloud,* on the other hand, pertains only to vocalizing a sound that is audible to others. So both of the following are **correct:**

"I'm sorry, but smoking is not allowed on school property."

"I'll whisper to you what I think about him, but I sure don't want to say it aloud."

all ready, already: A common confusion, but an easy one to correct. *Already* is used to indicate that something has happened previously or before, as in

"I already told you three times where I want to go to dinner."

Making two words out of it is simply a way of saying that all of those who are involved are ready. It refers to a group's preparedness for something yet to come, and it has nothing to do with what has happened previously:

"We're all ready to leave as soon as you finish packing the car."

allude, elude: *Allude* means to make a reference to; *elude* means to attempt avoidance of or escape from:

"I believe I alluded to that in my earlier comments."

"I think we're being followed by FBI agents. Let's try to elude them."

A lot, alot: Not a word! If you're talking about distribution, as in an "allotment of resources," then the correct word to use is *allot.* But you probably mean "a great many" of something. So either type exactly that, or type "a lot," with a proper space separation between the two words. That's what it has to be: two separate words.

[The writer of this section of the handbook had trouble, as well as embarrassment and confusion, making this item come out right. The culprit was his computer, which insisted on being overly helpful. The discovery came upon proofreading this very item and finding that it began: "a lot: Not a word!" Well, that doesn't make any sense at all. So what happened? The spell-checker function did its work automatically at a point where the writer didn't want it to do anything. That's what happened. What had been typed was "alot," but the computer recognized that as an illegal word and automatically corrected it! Now apart from the extra labor involved in understanding and repairing such a silly mess in text where one wants to show the wrong word, this is a good thing—in a way. But it's a bad thing, too, because it removes the opportunity for the writer to become aware of a writing failure and to learn how to do it properly. In effect, it rewards and encourages poor usage by refusing to divulge that poor usage to us. Since there may be countless other examples of computers' overassisting us in our writing, it is worthy of mention here—and worthy of our diligence as we genuinely seek to become better writers.]

anecdote, antidote: Far too often, we hear people who should know better say something like, "Let me tell you a little antidote about Sam." These folks must not realize that what they are actually saying is, "Let me tell you a little remedy that counteracts the effects of a poison about Sam." Because that is precisely what an antidote is: a remedy that counteracts the effects of a poison. What they mean to say, no doubt, is, "Let me tell you a little anecdote about Sam," an anecdote being an account of an interesting (and often humorous) incident. The distinction should be easy to remember, because the prefix *anti* always means against or in opposition to, for example, *antiaircraft, antibiotic, anticlimax*. In this case, we aren't generally against humorous stories, but we are generally against the various effects of poisoning. So **correct** usages include the following:

"I'd sure like to hear one of your clever anecdotes about classroom experiences."

"My God! The dog just drank some of the weed killer. Does the label say whether there's an antidote?"

(But if someone says, "Here's an antidote I think you'll enjoy," it's a little hard to tell which meaning he or she intends.)

appraise, apprize: *Appraise* means to evaluate. *Apprize* means to notify. Many people use the former when what they really mean is the latter. Here's a helpful hint: Just remember that *apprize* rhymes with *advise*. To advise someone is to apprize someone. (Well, not exactly, but close enough to work as a memory device.)

Wrong: "I'll appraise him of our progress."

Right: "The tax collector wants to appraise our house again!"

"Please allow me to apprize you of our condition."

bare, bear: One is naked. The other is a large animal that you usually don't want to fool around with. If you stripped the hair off of a grizzly, you'd have a bare bear. But *bear* can also mean carry, so watch out. If someone asked you to haul away the animal we just described, he or she would be asking you to bear bare bear. And to continue this silliness just a step further, if you were very tired that day and felt that you could hardly do it, you could answer: "I'm afraid I can barely bear bare bear." Next?

capital, capitol: No wonder this distinction is confusing. But it's pretty easy, once you learn the trick. *Capitol* means only the building in which a legislature meets. All other meanings (including the city that is the seat of government for a state or a country!) fall under *capital.* So,

> "The U.S. capitol building is located in Washington, D.C., the country's capital."

Also within the -tal spelling comes

1. money, wealth, assets

> "Once you acquire enough capital, you can start your own business."

2. involving death or calling for the death penalty

> "In most states, a first-degree murder conviction can result in capital punishment."

3. description of an upper-case letter of the alphabet

> "Generally speaking, the first word in a sentence should be capitalized."

choose, chose: This pair looks like it might be a relative of *loose* and *lose* (discussed later), so that the same rules and guidelines would apply. But thanks to the never-ending confusion and inconsistency of the English language, it's really entirely different. Maybe that's why people get a little mixed up. What is especially confusing is that the two that rhyme are *choose* and *lose,* which look like they shouldn't. Meanwhile, *chose* and *loose* don't rhyme with each other but do rhyme, respectively, with *nose* and *noose.* Go figure! Anyway, the main thing to know is that *choose* is for present and future tenses, while *chose* is for past tense.

> "I choose to ignore the comparison to *loose* and *lose;* what a dumb thing to tell us!"

> "I believe you chose the noose instead of the nose, you dummy. You'd better hope they tie it loose."

complement, compliment: *Complement* means to supplement, to fit harmoniously with; *compliment* means to say something nice, an expression of praise, admiration, or congratulation.

"She complimented him on his fine abilities as a seamstress."

"Your attention is the finest compliment I could possibly receive."

"This white wine would be the perfect complement to such a delicious dinner."

could/couldn't care less: In a strange but common speaking and writing error, people try to indicate extreme lack of interest or concern about something by saying, "I could care less." Ironically, this phrasing communicates exactly the opposite of the intent, almost like saying "Yes" when you really mean "No" because if you could care less, that implies that you do care currently and have a big range of lesser caring that you have not yet tapped into. So the right expression is "couldn't care less," meaning, "I have reached the lowest limit of my caring anything at all about it."

So "I couldn't care less what you think of my writing abilities" is **correct.**

But "I could care less about whether I communicate well" is **incorrect** (unless the speaker is trying to indicate that he or she does care quite a lot).

desert, dessert: One you eat after the main course has been completed. The other you die in if there's no water available because conditions are so insufferably hot and dry. (Well, my mother actually once made something to eat after the main course that was insufferably hot and dry. But that's another story.) Which is which, and how can a person remember? *Desert,* with the first syllable emphasized, is the hot, dry place. *Dessert,* with the second syllable emphasized, is the cake or pie or ice cream you eat after your meal. So that's one way to remember: The one with the *first* syllable emphasized has *one s,* and the one with the *second* syllable emphasized has *two.* If that isn't enough, remember that one time lost in the desert is enough, but seconds are sometimes nice for dessert.

"I want to finish eating my dessert before we continue our drive across the desert."

e.g., i.e.: Another frequently confused distinction. *I.e.* is an abbreviation for the Latin *id est,* meaning "that is." *E.g.* is an abbreviation for the Latin *exempli gratia,* meaning "for example." Use *i.e.* when you're trying to rephrase the same idea in different words. Use *e.g.* when you want to list one or more examples of whatever it was you just mentioned.

Correct: "Her message was succinct, i.e., brief and to the point."

"The package contained a variety of documents, e.g., notes, photographs, and maps."

Incorrect: "Her message was succinct, e.g., brief and to the point."

"The package contained a variety of documents, i.e., notes, photographs, and maps."

Note also that a period follows each letter in both abbreviations and that a comma always precedes the abbreviation and follows it (after the second period).

elicit, illicit: There's a big difference here, so be careful. To elicit something is to bring it out or call it forth, as in,

"The detective attempted to elicit details from the victim."

Something that is illicit, on the other hand, is improper or not sanctioned by custom or law as being proper or lawful, as in,

"The president and Ms. Lewinsky are alleged to have engaged in illicit sexual activity."

If it helps, *illicit* is an adjective (a modifier), while *elicit* is a verb.

eminent, imminent: Lots of people must not realize that these are two separate words. The most frequent error is in using *eminent* when what the writer/speaker really means is *imminent,* as in,

"The long awaited meeting is now eminent." **Wrong**

This should be,

"The long awaited meeting is now imminent,"

because *imminent* means that something is about to happen, whereas *eminent* usually refers to a person who is of special distinction of some sort. Hence,

"He is one of the most eminent geologists in the world." **Right**

ensure, insure: Good news. These two mean pretty much the same thing. Feel free to use them interchangeably, although *insure* has become the more commonly used of the pair. Both words mean to make secure or certain. So does a third word: *assure.* Perhaps the only difference worth noting, for those who want to be completely correct, is that *assure* is the most appropriate when referring to a person, as in the context of putting a person's mind at ease:

"I can assure you that I feel perfectly fine now."

The other two would most likely be found in a sentence such as,

"Putting some money aside now will help insure [or ensure] that we can pay the IRS when income tax time rolls around."

etc.: *Etc.,* not *ect.,* as it is commonly written. It is an abbreviation for the Latin *et cetera* (two words), meaning "and so forth."

formally, formerly: *Formerly* indicates something that happened in the past, whereas *formally* pertains to the opposite of casual or relaxed, whether the context is style of dress, furniture arrangement, structure of a term paper, and so forth.

> "I was formerly crude, rude, lewd, and unenlightened, but now I'm a sociology major."

> "The honors banquet is tomorrow night, and I would strongly advise you to dress formally for the occasion."

idle, idol: *Idle* means unoccupied, not busy, not in use. An idol, on the other hand, is an image of some sort that is used as an object of worship. In contemporary usage, *idol* has lost much of its original religious tone and is often used to refer to a person who is strongly admired. Hence, the following sentences are both **correct:**

> "Michael Jordan is my idol."

> "Since he retired from the NBA, Michael Jordan spends more of his time just being idle."

imply, infer: Very tricky. Often misused. Often used interchangeably, and they shouldn't be. Be careful here. To infer is to draw a conclusion, usually based on logical reasoning. To imply is to suggest or to express indirectly rather than directly. So implication lies within the speaker's remark, while inference is a conclusion made by someone else about the speaker's remark. Clear as mud? Try to sense the difference in the form of the following sentences, both of which are **correct.**

> "Do you mean to imply that my cooking is inferior?" [Focus is on the content of the other person's remark, who has just said something like, "I haven't eaten this well since my last trip to McDonald's."]

> "I infer from your comment that you don't care for my cooking." [Focus is on the cook's/speaker's interpretation of the crack about McDonald's.]

its, it's: There is considerable confusion about this distinction. The best way to remember the difference is to remember that *it's* is a contraction for *it is*. Although we almost always put an apostrophe before the *s* to indicate possession, that is a no-no in the case of these two words; when intended as a possessive, *its* does not contain an apostrophe. Hence, the following sentences are **correct:**

"It's about time we did something about improving student word usage."

"The paper suffered from its poor choice of word usage."

And the following sentences are **not correct:**

"The movement reached it's climax in 1983."

"Its very clear to me that this sentence is wrong."

lead, led: Here's another one that is commonly misused by college professors and others who should know better. If you're talking about the act of guiding someone or something in either the present or future tense, then *lead* (rhymes with *seed*) is the proper choice:

"The Pope will now lead us in prayer."

"Who's leading this group, anyway?"

If you're still talking about guidance but the guidance has already happened (past tense), then the correct word is *led* (rhymes with *bed*), not *lead:*

"She led us to the brink of disaster."

The confusion, of course, is that the stuff that makes a pencil write is pronounced the same way as *led* is pronounced.

Wrong, but common: "She lead us directly to our intended destination."

lose, loose: A mnemonic might help with this one. (A mnemonic is a device used as an aid to memory.) Q: What do you use to hang somebody? A: A noose. Q: What is it if it isn't tight? A: It's loose. So *loose* sounds like *noose* and means the opposite of tight. Meanwhile, *lose* (as in "If you don't stop with these silly examples, I'm afraid I'll lose my mind") rhymes with *accuse, abuse, dues, moos, sues,* and *twos.* Make up your own mnemonic!

nuclear, nucular: Not very many people actually spell this word wrong. It's just that a whole lot of otherwise fairly intelligent folks for some reason mispronounce it regularly—as "nucular," which isn't a real word and doesn't mean anything. The English language is a strange one, and it does contain a number of words that are pronounced differently than might appear proper. But this isn't one of them. When someone says "nucular," one response might be to say that it's "uncular" to you exactly what he or she means. When the person expresses puzzlement, explain that you're just trying to help out and be consistent. After all, shouldn't the -clear part that follows un- be pronounced the same way as the -clear part that follows nu-?

peace, piece: Not too difficult, but confused by some. A piece is a portion of something, as in, "I sure would like a piece of that apple pie." Peace is the condition of the absence of conflict. (There are many examples of how people who like to play with words have had fun with bumper stickers. "Visualize World Peace" has become, for the fun-with-words addict, "Visualize Whirled Peas.")

personal, personnel: Something that is personal is something that relates to a particular individual, that is, a particular person. *Personnel* refers to the group of individuals employed by a particular organization; it always pertains to the context of employment. **Correct:**

> "That score is a personal best for her."
>
> "Your question strikes me as being far too personal."
>
> "Most people conduct their personal hygiene rituals in the morning rather than in the evening."
>
> "This new rule applies to all personnel in the billing department of the company. Personnel in other divisions may ignore it."

plain, plane: Not too difficult, but occasionally confused. *Plain* means ordinary; *plane* means something you ride in up in the sky.

principal, principle: Some of us learned in childhood that the person who runs your school is the "princi*pal*" because he or she is your pal. OK, so what about the several other meanings? For the most part, *principle* has to do with things that are basic (including basic truths), and *principal* has to do with things that are supreme, first, or foremost.

> "This handbook is based on the principle that improvement in writing is a good thing."
>
> "The principal goal of this handbook is to improve the student's writing skills."

The Smith's, the Smiths, the Smiths's: We could use almost any last name, not just *Smith*, to make the point. Also, this is an error we're more likely to see on the name signs people put on their houses than in written documents. Nevertheless, a plural is constructed, in most cases, by simply adding an *s* to the word. So more than one Smith becomes *Smiths*.

If the sign on the house is intended to convey that more than one Smith lives here, it should say *The Smiths*. If the intent is to show that the property is owned by two or more members of the Smith family (possessive usage), the proper signage would be *The Smiths's House* (with the apostrophe and *s* following the *s* that indicates plural). Only if one Smith lives there alone and he or she wants to indicate ownership would the sign say *Smith's House,* and it would not have the word *The* in front of it.

But what you'll often see up and down the street is the **incorrect** *The Smith's.*

stationary, stationery: The one with an *e* is the one we write letters on. The one with an *a* means not moving or incapable of being moved. I can think of no simple tricks for remembering which is which, so please let me know if you come up with something that works. Meanwhile, these are **correct:**

> "It will be difficult to have discussion groups in this classroom because the seats are all stationary."
>
> "I received your note in the mail, and I must ask where you purchased such beautiful stationery."

tack, tact: An amazing number of professionals (including college professors, once again) misuse this one frequently. *Tack,* at least in this context, derives from the nautical setting and pertains to changing the course, or direction, of a vessel. Transferred to the interpersonal situation, the reference is still to changing course or direction, as in,

> "Well, that argument didn't work, so I think I'll take another tack."

Tact, on the other hand, refers to sensitivity to what is appropriate. It means much the same thing as *diplomacy:*

> "This is a very delicate situation. I'm going to have to use a great deal of tact."
>
> **Wrong,** but common: "Let's try a different tact."

their, there, they're: This is another pretty basic one, but a great many students (i.e., *a lot* of them) trip and fall over it all too frequently. *Their* is possessive:

> "Many students failed to visit with their advisors prior to registration."

There designates place:

> "We go there often."

There is also used to introduce a clause or sentence:

> "There were several excellent points made during the meeting."

See your dictionary for still other uses of *there.*

Finally, *they're* is a contraction of *they are:*

> "They're going to arrive in just a few minutes."

To, two, too: Pretty obvious, but be careful. Actually, the main problem seems to be with *to* versus *too. Too* means also or in excess, while *to* directs an action or destination. A mixed example of proper usage:

"I have already expressed this to you too many times. You, too, should understand it by now."

weather, whether: Whether you use *weather* or *whether* depends on whether you want to focus on atmospheric conditions or on alternative possibilities. So the following sentence is **correct** in both respects:

"I think today's weather is unbearably hot and humid, and I don't care whether you agree with me or not."

who's, whose: As usual, one is a contraction (in this case, for *who is*) and the other is a possessive. Which is which? One of the following pairs of sentences is right. The other is wrong. Can you tell which?

A: "I forgot who's turn it is."

"She's the one whose going to get us out of this mess."

B: "Who's sorry now?"

"Are you the person whose car is blocking mine?"

with (or in) regards to: Very common, and very irritating to those who know better. Some very esteemed people of lofty stature are guilty of this one. Whenever there's an *s* on the end of the word *regard,* it can refer to only two things. One of those is plural sentiments that are being expressed:

"Ken sends his kindest regards."

The other is the present tense of *to regard,* as in,

"He regards murder as a case of very bad manners."

Whenever the intended use of the word is as a synonym for *in reference to* or *in connection with,* it carries no letter *s* at the end.

Right: "With regard to your appearance, I find it beautiful."

"I am writing in regard to your recent advertisement."

Wrong: "With regards to this handbook, I find it worthless."

"I'd like to speak to you in regards to a raise in pay."

your, you're: Finally, this pair causes confusion with amazing frequency, including in such surprising places as newspaper headlines. Nine times out of 10, it is a case of the writer's using *your* when *you're* would actually be the proper choice. The apostrophe in *you're,* as usual, signals a contraction of two words into one; in this case, *you are* becomes *you're.* So if the statement could be made with equal accuracy by saying "you are," that means that *you're* is the proper choice. *Your,* on the other hand, refers to possession, for example, your book, your house, your relationship, your career. And it cannot (or shall we say should not) ever be used to mean you are.

Wrong: "Your going to regret this tomorrow."

"Your my favorite professor."

"Please leave you're shoes by the door."

Right: "I think you're going to like what I have to tell you."

"You may use your own pencil, if you wish."

Other Tidbits

Syntax. Syntax is the arrangement of words within a sentence. How we put the very same words together in a sentence can make a big difference in the clarity and the accuracy of what we are trying to express. Here's a three-word sentence with the three words arranged in all possible combinations. See which ones make sense and which ones don't. And (very important) see which totally different meanings can be discovered by comparing the ones that do make sense.

1. I here am.
2. I am here.
3. Am here I.
4. Am I here.
5. Here am I.
6. Here I am.

You should have found two that make no sense (1 and 3), one that makes sense but sounds oddly old-fashioned (5), one that would make sense if it had a different punctuation mark at the end (4), and two that are perfectly fine but that carry substantially different meanings from one another (2 and 6). If we can find such disparity in sentences made up of only three words (which we don't run into very often), imagine the confusion we can generate by sloppy, inattentive syntax in the longer sentences we write.

Try to think of examples of misuse of syntax. A fairly common one pertains to the context of whether all members of a category are alike (often heard in product commercials and everyday conversation):

Wrong: All students are not alike.

Right: Not all students are alike.

Students are not all alike.

Punctuation. Correct punctuation can be far more vitally important than most students realize. It is important not just because of tradition or because of some arbitrary academic standard of what is proper but because it can radically alter the meaning that the words convey. Here's an all-time favorite example. Note that the words are identical and even the syntax is identical. All that is changed is the punctuation. Are the two sentences equivalent in meaning?

Version A: "Woman, without her man, is nothing."

Version B: "Woman! Without her, man is nothing."

Index

About the Author

Since 1997, Mark Edwards (PhD, University of Washington) has taught stratification, research methods and writing, and social statistics in the sociology department at Oregon State University. His recent research on food insecurity has been used by nonprofit groups, state agencies, and the media to address domestic hunger in the Northwest. Many of his research papers have appeared in social science journals such as *Social Forces, Rural Sociology,* and *Social Science Quarterly*. But a favorite part of his work is helping students improve their social science writing.